Plastics Fabrication and Recycling

PLASTICS ENGINEERING

Founding Editor

Donald E. Hudgin

Professor
Clemson University
Clemson, South Carolina

1. Plastics Waste: Recovery of Economic Value, *Jacob Leidner*
2. Polyester Molding Compounds, *Robert Burns*
3. Carbon Black-Polymer Composites: The Physics of Electrically Conducting Composites, *edited by Enid Keil Sichel*
4. The Strength and Stiffness of Polymers, *edited by Anagnostis E. Zachariades and Roger S. Porter*
5. Selecting Thermoplastics for Engineering Applications, *Charles P. MacDermott*
6. Engineering with Rigid PVC: Processability and Applications, *edited by I. Luis Gomez*
7. Computer-Aided Design of Polymers and Composites, *D. H. Kaelble*
8. Engineering Thermoplastics: Properties and Applications, *edited by James M. Margolis*
9. Structural Foam: A Purchasing and Design Guide, *Bruce C. Wendle*
10. Plastics in Architecture: A Guide to Acrylic and Polycarbonate, *Ralph Montella*
11. Metal-Filled Polymers: Properties and Applications, *edited by Swapan K. Bhattacharya*
12. Plastics Technology Handbook, *Manas Chanda and Salil K. Roy*
13. Reaction Injection Molding Machinery and Processes, *F. Melvin Sweeney*
14. Practical Thermoforming: Principles and Applications, *John Florian*
15. Injection and Compression Molding Fundamentals, *edited by Avraam I. Isayev*
16. Polymer Mixing and Extrusion Technology, *Nicholas P. Cheremisinoff*
17. High Modulus Polymers: Approaches to Design and Development, *edited by Anagnostis E. Zachariades and Roger S. Porter*
18. Corrosion-Resistant Plastic Composites in Chemical Plant Design, *John H. Mallinson*
19. Handbook of Elastomers: New Developments and Technology, *edited by Anil K. Bhowmick and Howard L. Stephens*
20. Rubber Compounding: Principles, Materials, and Techniques, *Fred W. Barlow*

21. Thermoplastic Polymer Additives: Theory and Practice, *edited by John T. Lutz, Jr.*
22. Emulsion Polymer Technology, *Robert D. Athey, Jr.*
23. Mixing in Polymer Processing, *edited by Chris Rauwendaal*
24. Handbook of Polymer Synthesis, Parts A and B, *edited by Hans R. Kricheldorf*
25. Computational Modeling of Polymers, *edited by Jozef Bicerano*
26. Plastics Technology Handbook: Second Edition, Revised and Expanded, *Manas Chanda and Salil K. Roy*
27. Prediction of Polymer Properties, *Jozef Bicerano*
28. Ferroelectric Polymers: Chemistry, Physics, and Applications, *edited by Hari Singh Nalwa*
29. Degradable Polymers, Recycling, and Plastics Waste Management, *edited by Ann-Christine Albertsson and Samuel J. Huang*
30. Polymer Toughening, *edited by Charles B. Arends*
31. Handbook of Applied Polymer Processing Technology, *edited by Nicholas P. Cheremisinoff and Paul N. Cheremisinoff*
32. Diffusion in Polymers, *edited by P. Neogi*
33. Polymer Devolatilization, *edited by Ramon J. Albalak*
34. Anionic Polymerization: Principles and Practical Applications, *Henry L. Hsieh and Roderic P. Quirk*
35. Cationic Polymerizations: Mechanisms, Synthesis, and Applications, *edited by Krzysztof Matyjaszewski*
36. Polyimides: Fundamentals and Applications, *edited by Malay K. Ghosh and K. L. Mittal*
37. Thermoplastic Melt Rheology and Processing, *A. V. Shenoy and D. R. Saini*
38. Prediction of Polymer Properties: Second Edition, Revised and Expanded, *Jozef Bicerano*
39. Practical Thermoforming: Principles and Applications, Second Edition, Revised and Expanded, *John Florian*
40. Macromolecular Design of Polymeric Materials, *edited by Koichi Hatada, Tatsuki Kitayama, and Otto Vogl*
41. Handbook of Thermoplastics, *edited by Olagoke Olabisi*
42. Selecting Thermoplastics for Engineering Applications: Second Edition, Revised and Expanded, *Charles P. MacDermott and Aroon V. Shenoy*
43. Metallized Plastics, *edited by K. L. Mittal*
44. Oligomer Technology and Applications, *Constantin V. Uglea*
45. Electrical and Optical Polymer Systems: Fundamentals, Methods, and Applications, *edited by Donald L. Wise, Gary E. Wnek, Debra J. Trantolo, Thomas M. Cooper, and Joseph D. Gresser*
46. Structure and Properties of Multiphase Polymeric Materials, *edited by Takeo Araki, Qui Tran-Cong, and Mitsuhiro Shibayama*
47. Plastics Technology Handbook: Third Edition, Revised and Expanded, *Manas Chanda and Salil K. Roy*
48. Handbook of Radical Vinyl Polymerization, *edited by Munmaya K. Mishra and Yusef Yagci*

49. Photonic Polymer Systems: Fundamentals, Methods, and Applications, *edited by Donald L. Wise, Gary E. Wnek, Debra J. Trantolo, Thomas M. Cooper, and Joseph D. Gresser*
50. Handbook of Polymer Testing: Physical Methods, *edited by Roger Brown*
51. Handbook of Polypropylene and Polypropylene Composites, *edited by Harutun G. Karian*
52. Polymer Blends and Alloys, *edited by Gabriel O. Shonaike and George P. Simon*
53. Star and Hyperbranched Polymers, *edited by Munmaya K. Mishra and Shiro Kobayashi*
54. Practical Extrusion Blow Molding, *edited by Samuel L. Belcher*
55. Polymer Viscoelasticity: Stress and Strain in Practice, *Evaristo Riande, Ricardo Díaz-Calleja, Margarita G. Prolongo, Rosa M. Masegosa, and Catalina Salom*
56. Handbook of Polycarbonate Science and Technology, *edited by Donald G. LeGrand and John T. Bendler*
57. Handbook of Polyethylene: Structures, Properties, and Applications, *Andrew J. Peacock*
58. Polymer and Composite Rheology: Second Edition, Revised and Expanded, *Rakesh K. Gupta*
59. Handbook of Polyolefins: Second Edition, Revised and Expanded, *edited by Cornelia Vasile*
60. Polymer Modification: Principles, Techniques, and Applications, *edited by John J. Meister*
61. Handbook of Elastomers: Second Edition, Revised and Expanded, *edited by Anil K. Bhowmick and Howard L. Stephens*
62. Polymer Modifiers and Additives, *edited by John T. Lutz, Jr., and Richard F. Grossman*
63. Practical Injection Molding, *Bernie A. Olmsted and Martin E. Davis*
64. Thermosetting Polymers, *Jean-Pierre Pascault, Henry Sautereau, Jacques Verdu, and Roberto J. J. Williams*
65. Prediction of Polymer Properties: Third Edition, Revised and Expanded, *Jozef Bicerano*
66. Fundamentals of Polymer Engineering, *Anil Kumar and Rakesh K. Gupta*
67. Handbook of Polypropylene and Polymer, *Harutun Karian*
68. Handbook of Plastic Analysis, *Hubert Lobo and Jose Bonilla*
69. Computer-Aided Injection Mold Design and Manufacture, *J. Y. H. Fuh, Y. F. Zhang, A. Y. C. Nee, and M. W. Fu*
70. Handbook of Polymer Synthesis: Second Edition, *Hans R. Kricheldorf and Graham Swift*
71. Practical Guide to Injection Blow Molding, *Samuel L. Belcher*
72. Plastics Technology Handbook: Fourth Edition, *Manas Chanda and Salil K. Roy*
73. Industrial Polymers, Specialty Polymers, and Their Applications, *Manas Chanda and Salil K. Roy*
74. Plastics Fundamentals, Properties, and Testing, *Manas Chanda and Salil K. Roy*
75. Plastics Fabrication and Recycling, *Manas Chanda and Salil K. Roy*

Plastics Fabrication and Recycling

Manas Chanda
Salil K. Roy

CRC Press is an imprint of the
Taylor & Francis Group, an **informa** business

The material was previously published in *Plastics Technology Handbook, Fourth Edition* © Taylor & Francis, 2007.

CRC Press
Taylor & Francis Group
6000 Broken Sound Parkway NW, Suite 300
Boca Raton, FL 33487-2742

First issued in paperback 2019

ISBN-13: 978-1-4200-8062-9 (hbk)
ISBN-13: 978-0-367-38712-9 (pbk)

Library of Congress Cataloging-in-Publication Data

Chanda, Manas, 1940-
Plastics fabrication and recycling / Manas Chanda and Salil K. Roy.
p. cm. -- (Plastics engineering ; 75)
Includes bibliographical references and index.
ISBN 978-1-4200-8062-9 (alk. paper)
1. Plastics. 2. Plastics--Recycling. I. Roy, Salil K., 1939- II. Title. III. Series.

TP1120.C44 2008
668.4'12--dc22 2008011624

Visit the Taylor & Francis Web site at
http://www.taylorandfrancis.com

and the CRC Press Web site at
http://www.crcpress.com

Contents

Appendices

Preface

What makes plastics the most versatile of all materials is the ease with which they can be given any desired shape and form. Molding and fabrication processes, however, vary depending on the type of polymers to be processed and the shape and form of the end products to be made. Tooling for plastics processing defines the shape of the end product. A *mold* forms a complete three-dimensional part and is used in a number of fabrications processes, such as compression molding, injection molding, blow molding, thermoforming, and reaction injection molding (RIM), whereas a *die* is used to form two of the three dimensions of a plastic part—the third dimension, usually thickness or length, being controlled by other process variables—and is used in fabrication processes such as extrusion, pultrusion, and thermoforming. (Many plastics processes, however, do not differentiate between the terms *mold* and *die.*) Considering their critical importance in shaping the plastic products, Chapter 1 on fabrication processes begins with a discussion of different types of molds and dies.

Traditionally, plastics are divided into two broad categories, namely, *thermoplastic* and *thermosetting.* Thermoplastic resins, usually obtained as a granular polymer, can be repeatedly melted or solidified by heating or cooling—heat softens or melts the material so that it can be formed, and subsequent cooling then hardens or solidifies the material in the given shape. Thermoplastic resins are, therefore, usually molded by extrusion, injection, blow molding, and calendering processes as well as by thermoforming. On the other hand, thermosetting resins, which are usually supplied as a partially polymerized molding compound and are cross-linked or "cured" during the fabrication process, are usually shaped by compression molding and RIM. However, with the progress of technology, the demarcation between thermoplastic and thermoset processing has become less distinct so much so that thermoset processes have been developed which make use of the processing characteristics of thermoplastics, and modified machinery and molding compositions are made available to extend the economics of thermoplastic processing to thermosetting materials. All these molding processes are described in detail in the first part of Chapter 1.

Fiber-forming thermoplastic polymers are processed by spinning into filaments, which in turn are made into yarn, tow, roving, staple, and cord. The three principal types of spinning processes, namely, melt spinning, dry spinning, and wet spinning are also discussed in Chapter 1, with a focus on the physico-chemical factors of the respective processes and their effects on end–product qualities.

Besides the main fabrication processes cited above, there are many other processes that have been developed to serve specific needs, such as casting processes and reinforcing processes. Both thermosets and thermoplastics can be cast, for which the commonly used resins are acrylics, polystyrene, polyesters, poly(vinyl chloride), phenolics, and epoxies. Two basic types of casting are used in the plastics industry—simple casting and plastisol casting, while there are three variations of the latter, namely, dip casting, slush casting, and rotational casting. Reinforced plastics (RPs), in which a resin—thermosetting or thermoplastic—is combined with a reinforcing agent that can be fibrous, powdered, spherical, crystalline, or whisker and made of organic, metallic, or ceramic material, occupy a special place in the industry. A host of molding methods can be used for RPs, such as hand layup or contact molding, spray-up, matched metal molding, vacuum-bag molding, pressure-bag molding, continuous pultrusion, filament winding, pre-preg molding, RIM, structural RIM, and resin transfer molding. All these methods have been given due consideration in Chapter 1. Various types of fibrous reinforcements used for these RPs and their methods of manufacture are also discussed.

Foamed plastics, also referred to as cellular or expanded plastics, are widely used in making insulation frames, core materials for load-bearing structures, packaging materials, and cushioning materials. In view of their importance and wide diversity of forms and applications, a relative large space has been given to discussion of foaming processes, foaming agents, and other materials used in foamed polymers. Different types of cellular polymers, e.g., low-density, high-density, single-component, multicomponent, fiber-reinforced, and syntactic foams, are compared from structural point of view. Details of common industrial foams with focus on manufacturing processes are presented. These include polystyrene foams, polyolefin foams, polyurethane foams, foamed rubbers, foamed epoxies, urea-formaldehyde foams, silicone foams, phenolic foams, PVC foams, and syntactic foams.

The technology used for rubber processing and making rubber goods is quite different from that used for conventional thermoplastic and thermosetting resins. A separate section is therefore provided to present a fairly detailed account of rubber compounding and processing technology as practiced in the industry, including also the use of reclaimed rubber. The manufacturing technology of major rubber products is included.

Besides the main fabrication processes mentioned above, a wide variety of other techniques have been developed to diversify the use of polymers and to find many new uses. The more important of these methods and processes are described in the last part of Chapter 1. These include coating processes (e.g., fluidized bed coating, spray coating, and electrostatic coating), powder molding techniques (e.g., static molding, rotational molding, and centrifugal casting), adhesive bonding of plastics (e.g., solvent cementing, adhesive bonding, and plastics-specific miscellaneous methods), and plastics welding (e.g., hot-gas welding, fusion welding, friction welding, high-frequency welding, and ultrasonic welding).

Commercial techniques for decorating plastics are almost as varied as plastics themselves. Depending on end-use applications or market demands, virtually any desired effect or combination of effects, shading of tone, and degree of brightness can be imparted to flexible or rigid plastic products. The last section of Chapter 1 is dedicated to a discussion of *decoration of plastics*. The main topics chosen for discussion are various painting operations, printing processes (gravure printing, flexography, screen process printing, and pad printing), hot stamping, in-mold decorating, embossing, electroplating, and vacuum metallizing.

Beginning its journey from the womb of a mold or the face of a die, a plastic product, after its useful life, usually finds itself lodged in a landfill or discarded on the wayside. The property of high durability or permanence that is considered a valuable attribute of plastic in its applications becomes a distinct liability when the plastic is discarded. Plastics left lying around after use do not disappear from view and such post-consumer waste as foam cups, detergent bottles, and discarded film is a visual annoyance. All this is because plastics are not naturally biodegradable. However, to consider this as a detriment is a questionable argument. Rather, it may well be considered an advantage. This is borne out by the fact that recycling of plastics materials is now an important field in the plastics industry, not just an activity born under environmental pressure. The second chapter in the book presents an overview of several important aspects of plastics recycling and developments in the field. Some of the topics that are highlighted in this review are then elaborated in the subsequent sections of the chapter. In addition, waste recycling problems and possibilities relating to a number of common plastics are discussed.

This book was originally included in our well-known *Plastics Technology Handbook*. As this book deals only with fabrication processes and recycling methods, it will be found to be informative and useful by those who are more interested in learning about the fabrication processes used for polymers and the various methods presently available for polymer recycling. To a large extent, the book owes its origin to the efforts of Allison Shatkin, who first conceived the idea of bringing out spin-off plastics books for the benefit of different groups of readers and also took the initiative in realizing them. We greatly appreciate her efforts.

Manas Chanda
Salil K. Roy

Authors

Manas Chanda has been a professor and is presently an emeritus professor in the Department of Chemical Engineering, Indian Institute of Science, Bangalore, India. He also worked as a summer-term visiting professor at the University of Waterloo, Ontario, Canada with regular summer visits from 1980 to 2000. A five-time recipient of the International Scientific Exchange Award from the Natural Sciences and Engineering Research Council, Canada, Dr. Chanda is the author or coauthor of nearly 100 scientific papers, articles, and books, including *Introduction to Polymer Science and Chemistry* (CRC Press/Taylor & Francis). A fellow of the Indian National Academy of Engineers and a member of the Indian Plastics Institute, he received a BS (1959) and MSc (1962) from Calcutta University, and a PhD (1966) from the Indian Institute of Science, Bangalore, India.

Salil K. Roy is a professor in the Postgraduate Program in Civil Engineering of the Petra Christian University, Surabaya, Indonesia. Earlier he worked as lecturer, senior lecturer, and associate professor at the National University of Singapore. Prior to that he was a research scientist at American Standard, Piscataway, New Jersey.

Dr. Roy is a fellow of the Institution of Diagnostic Engineers, U.K., and has published over 250 technical papers in professional journals and conference proceedings; he also holds several U.S. Patents. He received a BSc (1958) and MSc (Tech.) (1961) from the University of Calcutta, India, and a ScD (1966) from the Massachusetts Institute of Technology, Cambridge, Massachusetts. Dr. Roy is a subject of biographical record in the prestigious *Great Minds of the 21st Century* published by the American Biographical Institute, *Who's Who in the World* published by the Marquis Who's Who in the World, and *2000 Outstanding Intellectuals of the 21st Century* published by the International Biographical Centre, Cambridge, England.

1

Fabrication Processes

1.1 Types of Processes

As indicated in Chapter 1 of *Plastics Fundamentals, Properties, and Testing*, the family of polymers is extraordinarily large and varied. There are, however, some fairly broad and basic approaches that can be followed when designing or fabricating a product out of polymers or, more commonly, polymers compounded with other ingredients. The type of fabrication process to be adopted depends on the properties and characteristics of the polymer and on the shape and form of the final product.

In the broad classification of plastics there are two generally accepted categories: thermoplastic resins and thermosetting resins.

Thermoplastic resins consist of long polymer molecules, each of which may or may not have side chains or groups. The side chains or groups, if present, are not linked to other polymer molecules (i.e., are not cross-linked). Thermoplastic resins, usually obtained as a granular polymer, can therefore be repeatedly melted or solidified by heating or cooling. Heat softens or melts the material so that it can be formed; subsequent cooling then hardens or solidifies the material in the given shape. No chemical change usually takes place during this shaping process.

In thermosetting resins the reactive groups of the molecules from cross-links between the molecules during the fabrication process. The cross-linked or "cured" material cannot be softened by heating. Thermoset materials are usually supplied as a partially polymerized molding compound or as a liquid monomer–polymer mixture. In this uncured condition they can be shaped with or without pressure and polymerized to the cured state with chemicals or heat.

With the progress of technology the demarcation between thermoplastic and thermoset processing has become less distinct. For thermosets processes have been developed which make use of the economic processing characteristics of thermoplastics. For example, cross-linked polyethylene wire coating is made by extruding the thermoplastic polyethylene, which is then cross-linked (either chemically or by irradiation) to form what is actually a thermoset material that cannot be melted again by heating. More recently, modified machinery and molding compositions have become available to provide the economics of thermoplastic processing to thermosetting materials. Injection molding of phenolics and other thermosetting materials are such examples. Nevertheless, it is still a widespread practice in industry to distinguish between thermoplastic and thermosetting resins.

Compression and transfer molding are the most common methods of processing thermosetting plastics. For thermoplastics, the more important processing techniques are extrusion, injection, blow molding, and calendaring; other processes are thermoforming, slush molding, and spinning.

1.2 Tooling for Plastics Processing

Tooling for plastics processing defines the shape of the part. It falls into two major categories, molds and *dies*. A mold is used to form a complete three-dimensional plastic part. The plastics processes that use molds are compression molding, injection molding, blow molding, thermoforming, and reaction injection molding (RIM). A die, on the other hand, is used to form two of the three dimensions of a plastic part. The third dimension, usually thickness or length, is controlled by other process variables. The plastics processes that use dies are extrusion, pultrusion and thermoforming. Many plastics processes do not differentiate between the terms mold and die. Molds, however, are the most predominant form of plastics tooling.

1.2.1 Types of Mold

The basic types of mold, regardless of whether they are compression, injection, transfer, or even blow molds, are usually classified by the type and number of cavities they have. For example, Figure 1.1 illustrates three mold types: (a) single-cavity, (b) dedicated multiple-cavity, and (c) family multiple-cavity.

Single-cavity mold (Figure 1.1a) represents one of the simplest mold concepts. This design lends itself to low-volume production and to large plastic part designs. The multiple-cavity molds may be of two types. A dedicated multiple-cavity mold (Figure 1.1b) has cavities that produce the same part. This type of mold is very popular because it is easy to balance the plastic flow and establish a controlled process. In a family multiple-cavity mold (Figure 1.1c), each cavity may produce a different part. Historically, family mold designs were avoided because of difficulty in filling uniformly; however, recent advances in mold making and gating technology make family molds appealing. This is the case especially when a processor has a multiple-part assembly and would like to keep inventories balanced.

1.2.2 Types of Dies

Within the plastics industry, the term *die* is most often applied to the processes of extrusion (see EXTRUSION). Extrusion dies may be categorized by the type of product being produced (e.g., film, sheet, profile, or coextrusion), but they all have some common features as described below.

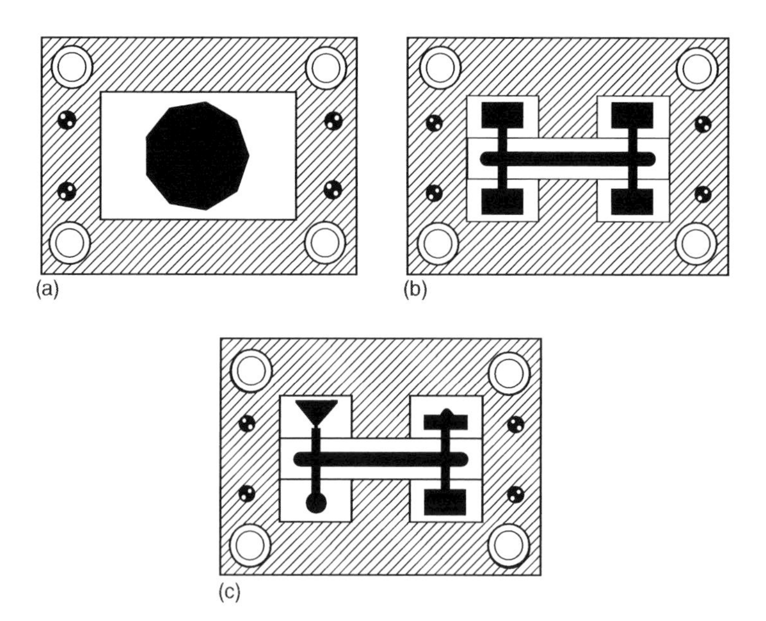

FIGURE 1.1 Three basic types of molds. (a) single-cavity; (b) dedicated multiple-cavity; (c) family multiple-cavity.

1. *Steel.* The extrusion process being continuous, both erosion and corrosion are significant factors. Hence the dies must be made of a high-quality tool steel, hardened so that the areas that contact the plastic material do not erode. Additionally, many dies have a dense, hard chrome plating in the area where plastic melt contacts the die.
2. *Heaters.* Extrusion dies are to be heated in order to maintain a melt flow condition for the plastic material. Most of the heaters are cartridge-type elements that slip fit into the die at particular locations. In addition to the heaters, the dies have to accommodate temperature sensors, such as thermocouples.
3. *Melt Pressure.* Many sophisticated dies are equipped with sensors that monitor melt pressure. This allows the processor to better monitor and control the process.
4. *Parting Line.* Large extrusion dies must be able to separate at the melt flow line for easier fabrication and maintenance. Smaller extrusion dies may not have a parting area, because they can be constructed in one piece.
5. *Die Swell Compensation.* The polymer melt swells when it exits the die, as explained previously. This die swell is a function of the type of plastic material, the melt temperature, the melt pressure, and the die configuration. The die must be compensated for die swell so that the extruded part has the corrected shape and dimensions. Molds and dies for different fabrication processes will be described later in more detail when the processes are discussed.

1.2.3 Tool Design

The design of the tooling to produce a specific plastics part must be considered during the design of the part itself. The tool designer must consider several factors that may affect the fabricated part, such as the plastics material, shrinkage, and process equipment. Additionally, competitive pressures within the plastics industry require the tool designer to consider how to facilitate tool changeovers, optimize tool maintenance, and simplify (or eliminate) secondary operations.

Historically, plastics molds and dies were built by toolmakers who spent their lives learning and perfecting their craft. Today the void created by the waning numbers of these classically trained toolmakers is being filled by the development of numerically controlled (NC) machinery centers, computer-based numerically controlled (CNC) machinery centers, and computer-aided design (CAD) systems. Molds and dies can now be machined on computer-controlled mills, lathes, and electric discharge machines that require understanding of computers and design, rather than years of experience and machining skills. The quality of tool components is now more a function of the equipment than of the toolmaker's skill.

The high costs of molds and the fact that many production molds are built under extreme time constraints leave no room for trial and error. Though prototyping has been widely used to evaluate smaller part designs when circumstances and time allow, prototyping is not always feasible for larger part designs. There are, however, several alternatives to prototyping, e.g., CAD, finite-element analysis (FEA), and rapid prototyping. While CAD allows a tool designer to work with a three-dimensional computer model of the tool being designed and to analyze the design, FEA allows the tool to be evaluated (on a computer) for production worthiness. The mold is then fabricated from the computer model, a process called computer-aided manufacturing (CAM).

Rapid prototyping is a relatively new method of producing a plastics part by using a three-dimensional computer drawing. A sophisticated prototyping apparatus interprets the drawing and guides an articulating laser beam across a specific medium such as a photopolymer plastic or laminated paper, the result being a physical representation of the computer-based drawing. Prototyped parts can be produced in less than 24 h, and part designs can be scaled to fit the size of the prototyping equipment. Another trend is the introduction of molds that accept interchangeable modules. Modules take less time to manufacturing, and in turn, cut down on the delivery time and costs. In addition, it usually takes less time to change the module than the entire mold frame.

1.3 Compression Molding

Compression molding is the most common method by which thermosetting plastics are molded [1–3]. In this method the plastic, in the form of powder, pellet, or disc, is dried by heating and then further heated to near the curing temperature; this heated charge is loaded directly into the mold cavity. The temperature of the mold cavity is held at 150°C–200°C, depending on the material. The mold is then partially closed, and the plastic, which is liquefied by the heat and the exerted pressure, flows into the recess of the mold. At this stage the mold is fully closed, and the flow and cure of the plastic are complete. Finally, the mold is opened, and the completely cured molded part is ejected.

Compression-molding equipment consists of a matched mold, a means of heating the plastic and the mold, and some method of exerting force on the mold halves. For severe molding conditions molds are usually made of various grades of tool steel. Most are polished to improve material flow and overall part quality. Brass, mild steel, or plastics are used as mold materials for less severe molding conditions or short-run products.

In compression molding a pressure of 2,250 psi (158 kg/cm^2)–3,000 psi (211 kg/cm^2) is suitable for phenolic materials. The lower pressure is adequate only for an easy-flow materials and a simple uncomplicated shallow molded shape. For a medium-flow material and where there are average-sized recesses, cores, shapes, and pins in the molding cavity, a pressure of 3,000 psi (211 kg/cm^2) or above is required. For molding urea and melamine materials, pressures of approximately one and one-half times that needed for phenolic material are necessary.

The time required to harden thermosetting materials is commonly referred to as the cure time. Depending on the type of molding material, preheating temperature, and the thickness of the molded article, the cure time may range from seconds to several minutes.

In compression molding of thermosets the mold remains hot throughout the entire cycle; as soon as a molded part is ejected, a new charge of molding powder can be introduced. On the other hand, unlike thermosets, thermoplastics must be cooled to harden. So before a molded part is ejected, the entire mold must be cooled, and as a result, the process of compression molding is quite slow with thermoplastics. Compression molding is thus commonly used for thermosetting plastics such as phenolics, urea, melamine, and alkyds; it is not ordinarily used for thermoplastics. However, in special cases, such as when extreme accuracy is needed, thermoplastics are also compression molded. One example is the phonograph records of vinyl and styrene thermoplastics; extreme accuracy is needed for proper sound reproduction. Compression molding is ideal for such products as electrical switch gear and other electrical parts, plastic dinnerware, radio and television cabinets, furniture drawers, buttons, knobs, handles, etc.

Like the molding process itself, compression molding machinery is relatively simple. Most compression presses consist of two platens that close together, applying heat and pressure to the material inside a mold. The majority of the presses are hydraulically operated with plateau ranging in size from 6 in. square to 8 ft square or more. The platens exert pressures ranging from 6 up to 10,000 tons. Virtually all compression molding presses are of vertical design. Most presses having tonnages under 1000 are upward-acting, while most over 1,000 tons act downward. Some presses are built with a shuttle-clamp arrangement that moves the mold out of the clamp section to facilitate setup and part removal.

Compression molds can be divided into hand molds, semiautomatic molds, and automatic molds. The design of any of these molds must allow venting to provide for escape of steam, gas, or air produced during the operation. After the initial application of pressure the usual practice is to open the mold slightly to release the gases. This procedure is known as breathing.

Hand molds are used primarily for experimental runs, for small production items, or for molding articles which, because of complexity of shape, require dismantling of mold sections to release them. Semiautomatic molds consist of units mounted firmly on the top and bottom platens of the press. The operation of the press closes and opens the mold and actuates the ejector system for removal of the molded article. However, an operator must load the molding material, actuate press controls for the molding sequence, and remove the ejected piece from the mold. This method is widely used.

Fully automatic molds are specially designed for adaptation to a completely automatic press. The entire operation cycle, including loading and unloading of the mold, is performed automatically, and all molding operations are accurately controlled. Thermosetting polymers can be molded at rates up to 450 cycles/h. Tooling must be of the highest standard to meet the exacting demands of high-speed production. Automatic molds offer the most economical method for long production runs because labor costs are kept to a minimum.

The three common types of mold designs are open flash, fully positive, and *semipositive.*

1.3.1 Open Flash

In an open flash mold a slight excess of molding powder is loaded into the mold cavity (Figure 1.1a) [4]. On closing the top and bottom platens, the excess material is forced out and flash is formed. The flash blocks the plastic remaining in the cavity and causes the mold plunger to exert pressure on it. Gas or air can be trapped by closing the mold too quickly, and finely powdered material can be splashed out of the mold. However, if closing is done carefully, the open flash mold is a simple one, giving very good results.

Since the only pressure on the material remaining in the flash mold when it is closed results from the high viscosity of the melt which did not allow it to escape, only resins having high melt viscosities can be molded by this process. Since most rubbers have high melt viscosities, the flash mold is widely used for producing gaskets and grommets, tub and flash stoppers, shoe heels, door mats, and many other items.

Because of lower pressure exerted on the plastic in the flash molds, the molded products are usually less dense than when made using other molds. Moreover, because of the excess material loading needed, the process is somewhat wasteful as far as raw materials are concerned. However, the process has the advantage that the molds are cheap, and very slight labor costs are necessary in weighing out the powder.

1.3.2 Fully Positive

In the fully positive molds (Figure 1.2b) no allowance is made for placing excess powder in the cavity [4]. If excess powder is loaded, the mold will not close; an insufficient charge will result in reduced thickness of the molded article. A correctly measured charge must therefore be used with this mold—it is a disadvantage of the positive mold. Another disadvantage is that the gases liberated during the chemical curing reaction are trapped inside and may show as blisters on the molded surface. Excessive wear on the sliding fit surface on the top and bottom forces and the difficulty of ejecting the molding are other reasons for discarding this type of mold. The mold is used on a small scale for molding thermosets, laminated plastics, and certain rubber components.

1.3.3 Semipositive

The semipositive mold (Figure 1.2c and d) combines certain features of the open flash and fully positive molds and makes allowance for excess powder and flash [4]. It is also possible to get both horizontal and vertical flash. Semipositive molds are more expensive to manufacture and maintain than the other types, but they are much better from an applications point of view. Satisfactory operation of semipositive molds is obtained by having clearance (0.025/25 mm of diameter) between the plunger (top force) and the cavity. Moreover, the mold is given a 2–3° taper on each side. This allows the flash to flow on and the entrapped gases to escape along with it, thereby producing a clean, blemish-free mold component.

1.3.4 Process Applicability

Compression molding is most cost-effective when used for short-run parts requiring close tolerances, high-impact strength, and low mold shrinkage. Old as the process may be, new applications continue to

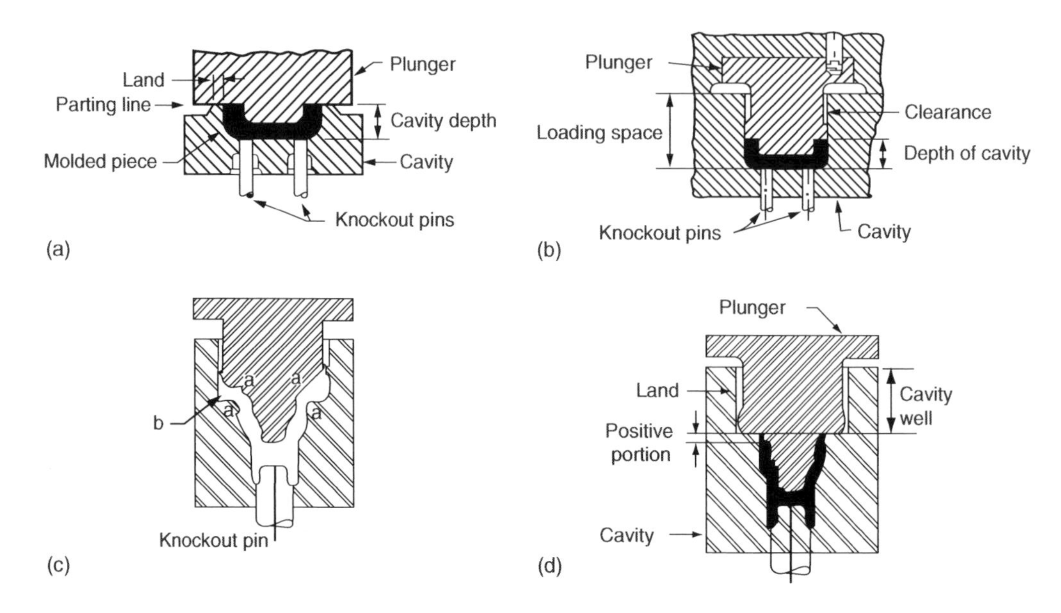

FIGURE 1.2 Compression molds. (a) A simple flash mold. (b) A positive mold. Knockout pins could extend through plunger instead of through cavity. (c) Semi-positive mold as it appears in partly closed position before it becomes positive. Material trapped in area b escapes upward. (d) Semipositive mold in closed position.

evolve compression molding. For example, in the dental and medical fields, orthodontic retainers, and pacemaker casings are now mostly compression molded because of low tool costs. Injection molding tools to produce the same part would cost as much as eight times more. Manufacturers of gaskets and seals who started out with injection-molded products to take advantage of the faster cycle times, are now switching back to compression molding to maintain quality level required for these parts.

The use of compression molding has expanded significantly in recent years due to the development of new materials, reinforced materials in particular. Molding reinforced plastics (RPs) requires two matched dies usually made of inexpensive aluminum, plastics, or steel and used on short runs.

Adding vacuum chambers to compression molding equipment in recent years has reduced the number of defects caused by trapped air or water in the molding compound, resulting in higher-quality finished parts. Another relatively new improvement has been the addition of various forms of automation to the process. For example, robots are used both to install inserts and remove finished parts.

1.4 Transfer Molding

In transfer molding, the thermosetting molding powder is placed in a chamber or pot outside the molding cavity and subjected to heat and pressure to liquefy it [1–6]. When liquid enough to start flowing, the material is forced by the pressure into the molding cavity, either by a direct sprue or though a system of runners and gates. The material sets hard to the cavity shape after a certain time (cure time) has elapsed. When the mold is disassembled, the molded part is pushed out of the mold by ejector pins, which operate automatically.

Figure 1.3 shows the molding cycle of *pot-type transfer* molding, and Figure 1.4 shows plunger-type transfer molding (sometime called auxiliary raw transfer molding). The taper of the sprue is pot-type transfer is such that, when the mold is opened, the sprue remains attached to the disc of material left in the pot, known as *cull*, and is thus pulled away from the molded part, whereas the latter is lifted out of the cavity by the ejector pins (Figure 1.3c). In plunger-type transfer molding, on the other hand, the cull and the sprue remain with the molded piece when the mold is opened (Figure 1.4c).

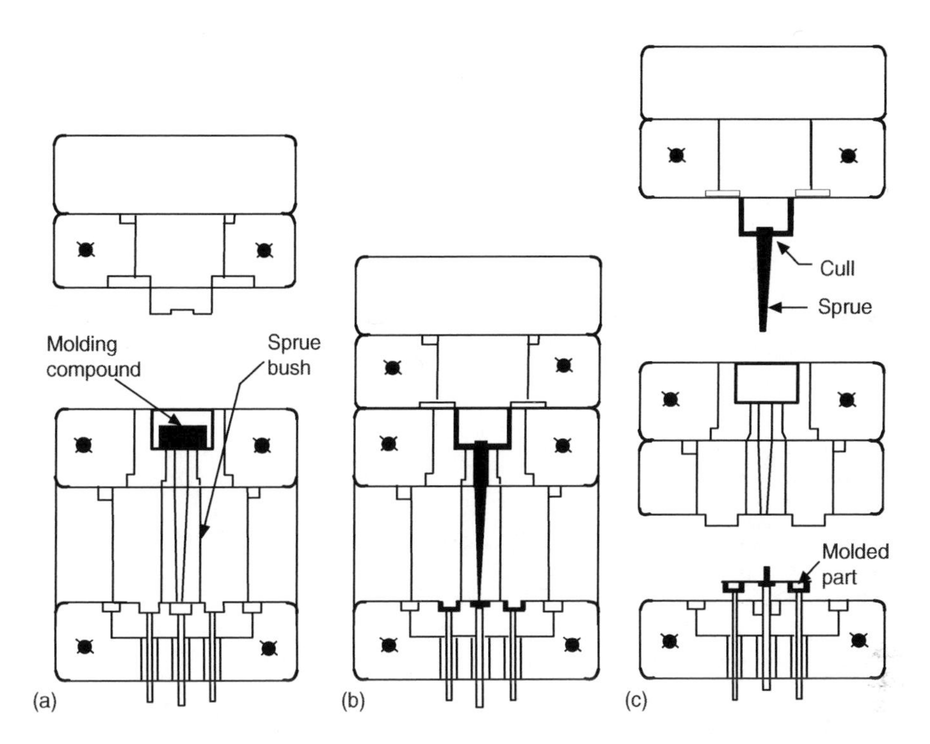

FIGURE 1.3 Molding cycle of a pot-type transfer mold. (a) Molding compound is placed in the transfer pot and then (b) forced under pressure when hot through an orifice and into a closed mold. (c) When the mold opens, the sprue remains with the cull in the pot, and the molded part is lifted out of the cavity by ejector pins. (After Vaill, E.W. September 1962. *Mod. Plastics*, 40, 1A, Encycl. Issue, 767.)

Another variation of transfer molding is screw transfer molding (Figure 1.5). In this process the molding material is preheated and plasticized in a screw chamber and dropped into the pot of an inverted plunger mold. The preheated molding material is then transferred into the mold cavity by the same method as shown in Figure 1.4. The screw-transfer-molding technique is well suited to fully automatic operation. The optimum temperature of a phenolic mold charge is 240°F $\pm$ 20°F (155°C $\pm$ 11°C), the same as that for pot-transfer and plunger molding techniques.

For transfer molding, generally pressures of three times the magnitude of those required for compression molding are required. For example, usually a pressure of 9,000 psi (632 kg/cm^2) and upward is required for phenolic molding material (the pressure referred to here is that applied to the powder material in the transfer chamber).

The principle of transferring the liquefied thermosetting material from the transfer chamber into the molding cavity is similar to that of the injection molding of thermoplastics (described later). Therefore the same principle must be employed for working out the maximum area which can be molded—that is, the projected area of the molding multiplied by the pressure generated by the material inside the cavity must be less than the force holding the two halves together. Otherwise, the molding cavity plates will open as the closing force is overcome.

Transfer molding has an advantage over compression molding in that the molding powder is fluid when it enters the mold cavity. The process therefore enables production of intricate parts and molding around thin pins and metal inserts (such as an electrical lug). Thus, by transfer molding, metal inserts can be molded into the component in predetermined positions held by thin pins, which would, however, bend or break under compression-molding conditions. Typical articles made by the transfer molding process are terminal-bloc insulators with many metal inserts and intricate shapes, such as cups and caps for cosmetic bottles.

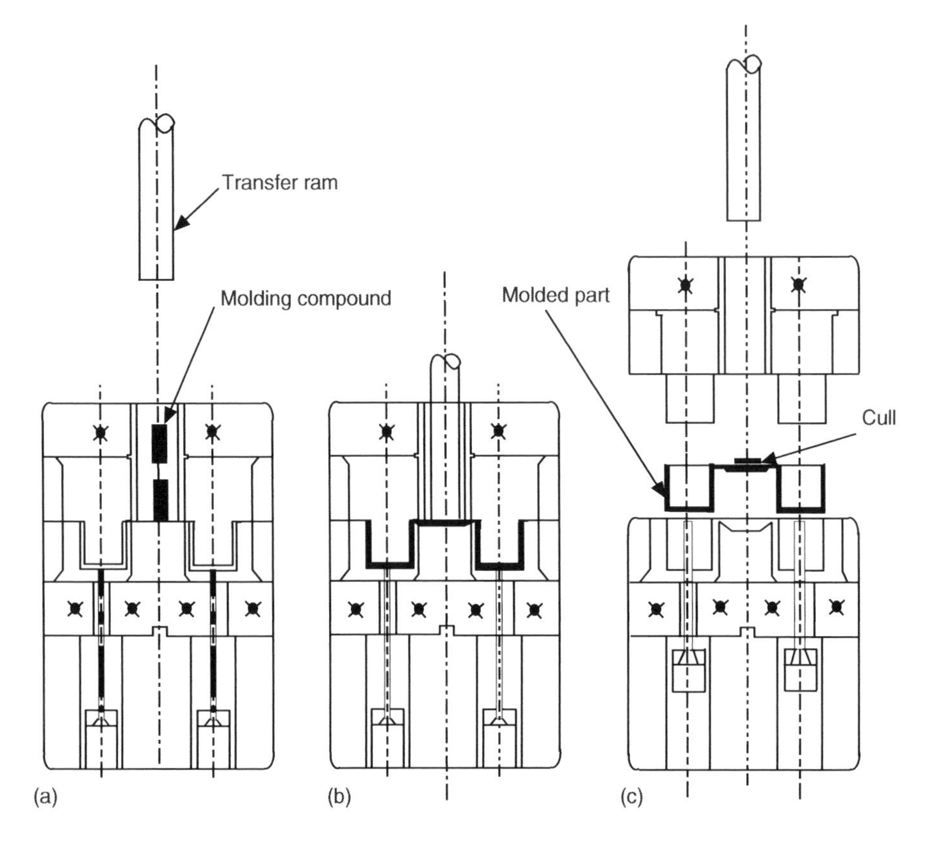

FIGURE 1.4 Molding cycle of a plunger-type transfer mold. (a) An auxiliary ram exerts pressure on the heat-softened material in the pot and (b) forces it into the mold. (c) When the mold is opened, the cull and sprue remain with the molded piece. (After Vaill, E.W. September 1962. *Mod. Plastics*, 40, 1A, Encycl. Issue, 767.)

1.4.1 Ejection of Molding

Ejection of a molded plastic article from a mold can be achieved by using ejector pins, sleeves, or stripper plates. Ejector pins are the most commonly used method because they can be easily fitted and replaced. The ejector pins must be located in position where they will eject the article efficiently without causing distortion of the part. They are worked by a common ejector plate or a bar located under the mold, and operated by a central hydraulic ejector ram. The ejector pins are fitted either to the bottom force or to the top force depending on whether it is necessary for the molding to remain in the bottom half of the female part or on the top half of the male part of the tool. The pins are usually constructed of a hardened steel to avoid wear.

1.4.2 Heating System

Heating is extremely important in plastics molding operations because the tool and auxiliary parts must be heated to the required temperature, depending on the powder being molded, and the temperature must be maintained throughout the molding cycle. The molds are heated by steam, hot waters, and induction heaters. Steam heating is preferred for compression and transfer molding, although electricity is also used because it is cleaner and has low installation costs. The main disadvantage of the latter method is that the heating is not fully even, and there is tendency to form hot spots.

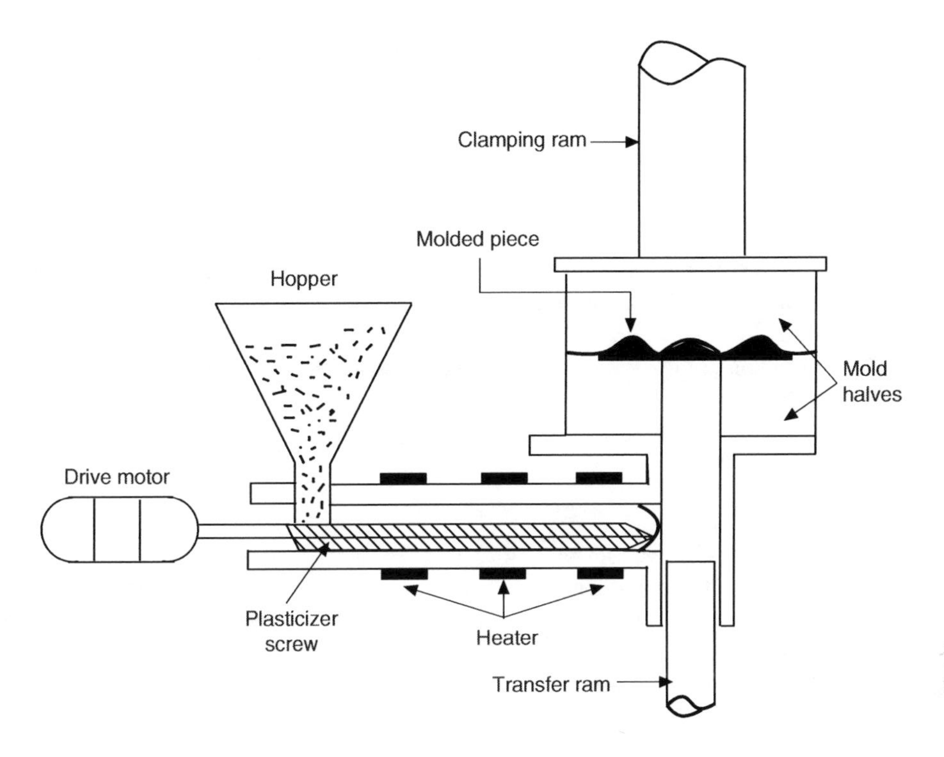

FIGURE 1.5 Drawing of a screw-transfer molding machine.

1.4.3 Types of Presses

Presses used for compression and transfer molding of thermosets can be of many shapes and designs, but they can be broadly classified as hand, mechanical, or hydraulic types. Hand presses have relatively lower capacity, ranging from 10 to 100 tons, whereas hydraulic presses have considerably higher capacity (500 tons). Hydraulic presses may be of the upstroke or downstroke varieties. In the simple upstroke press, pressure can be applied fairly quickly, but the return is slow. In the downstroke press fitted with a prefilling tank, this disadvantage of the upstroke press is removed, and a higher pressure is maintained by prefilling with liquid from a tank.

The basic principles of hydraulics are used in the presses. Water or oil is used as the main fluid. Water is cheap but rusts moving parts. Oil is more expensive but it does not corrode and it does lubricate moving parts. The main disadvantage of oil is that it tends to form sludge due to oxidation with air.

The drive for the presses is provided by single pumps or by central pumping stations, and accumulators are used for storing energy to meet instantaneous pressure demand in excess of the pump delivery. The usual accumulator consists of a single-acting plunger working in a cylinder. The two main types of accumulators used are the weight-loaded type and the air-loaded type. The weight-loaded type is heavy and therefore not very portable. There is also an initial pressure surge on opening the valve. The pressure-surge problem is overcome in the air- or gas (nitrogen)-loaded accumulator. This type is more portable but suffers a small pressure loss during the molding cycle.

1.4.4 Preheating

To cut down cycle times and to improve the finished product of compression molding and transfer molding, the processes of preheating and performing are commonly used. With preheating, relatively thick sections can be molded without porosity. Other advantages of the technique include improved flow of resin, lower molding pressures, reduced mold shrinkage, and reduced flash.

Preheating methods are convection, infrared, radio frequency, and steam. Thermostatically controlled gas or electrically heated ovens are inexpensive methods of heating. The quickest, and possibly the most efficient, method is radio-frequency heating, but it is also the most expensive. Preheaters are located adjacent to the molding press and are manually operated for each cycle.

1.4.5 Preforming

Preforming refers to the process of compressing the molding powder into the shape of the mold before placing it in the mold or to pelleting, which consists of compacting the molding powder into pellets of uniform size and approximately known weight. Preforming has many advantages, which include avoiding waste, reduction in bulk factor, rapid loading of charge, and less pressure than uncompacted material. Preformers are basically compacting presses. These presses may be mechanical, hydraulic, pneumatic, or rotary cam machines.

1.4.6 Flash Removal

Although mold design takes into consideration the fact that flash must be reduced to a minimum, it still occurs to some extent on the molded parts. It is thus necessary to remove the flash subsequent to molding. This removal is most often accomplished with tumbling machines. These machines tumble molded parts against each other to break off the flash. The simplest tumbling machines are merely wire baskets driven by an electric motor with a pulley belt. In more elaborate machines blasting of molded parts is also performed during the tumbling operation.

1.5 Injection Molding of Thermoplastics

Injection molding is the most important molding method for thermoplastics [7–9]. It is based on the ability of thermoplastic materials to be softened by heat and to harden when cooled. The process thus consists essentially of softening the material in a heated cylinder and injecting it under pressure into the mold cavity, where it hardens by cooling. Each step is carried out in a separate zone of the same apparatus in the cyclic operation.

A diagram of a typical injection-molding machine is shown in Figure 1.6. Granular material (the plastic resin) falls from the hopper into the barrel when the plunger is withdrawn. The plunger then pushes the material into the heating zone, where it is heated and softened (plasticized or plasticated). Rapid heating takes place due to spreading of the polymer into a thin film around a torpedo. The already molten polymer displaced by this new material is pushed forward through the nozzle, which is in intimate contact with the mold. The molten polymer flows through the sprue opening in the die, down the runner, past the gate, and into the mold cavity. The mold is held tightly closed by the clamping action of the press platen. The molten polymer is thus forced into all parts of the mold cavities, giving a perfect reproduction of the mold.

The material in the mold must be cooled under pressure below T_m or T_g before the mold is opened and the molded part is ejected. The plunger is then withdrawn, a fresh charge of material drops down, the mold is closed under a locking force, and the entire cycle is repeated. Mold pressures of 8,000–40,000 psi (562–2,812 kg/cm^2) and cycle times as low as 15 sec are achieved on some machines.

Note that the feed mechanism of the injection molding machine is activated by the plunger stroke. The function of the torpedo in the heating zone is to spread the polymer melt into thin film in close contact with the heated cylinder walls. The fins, which keep the torpedo centered, also conduct heat from the cylinder walls to the torpedo, although in some machines the torpedo is heated separately.

Injection-molding machines are rated by their capacity to mold polystyrene in a single shot. Thus a 2-oz machine can melt and push 2 oz of general-purpose polystyrene into a mold in one shot. This capacity is determined by a number of factors such as plunger diameter, plunger travel, and heating capacity.

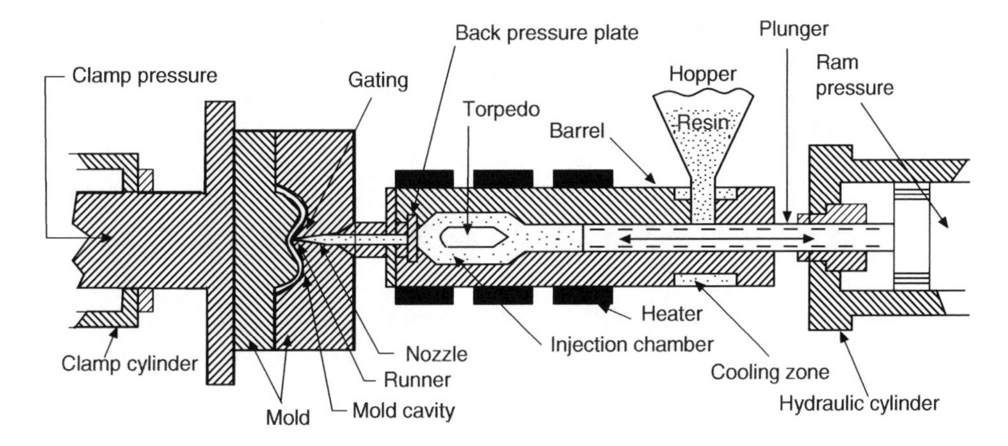

FIGURE 1.6 Cross-section of a typical plunger injection-molding machine. (After *Petrothene: A Processing Guide*, 3rd Ed., 1965. U.S. Industrial Chemicals Co., New York.)

The main component of an injection-molding machine are (1) the injection unit which melts the molding material and forces it into the mold; (2) the clamping unit which opens the mold and closes it under pressure; (3) the mold used; and (4) the machine controls.

1.5.1 Types of Injection Units

Injection-molding machines are known by the type of injection unit used in them. The oldest type is the single-stage plunger unit (Figure 1.6) described above. As the plastic industry developed, another type of plunger machine appeared, known as a two-stage plunger (Figure 1.7a). It has two plunger units set one on top of the other. The upper one, also known as a preplasticizer, plasticizes the molding material and feeds it to the cylinder containing the second plunger, which operates mainly as a shooting plunger, and pushes the plasticized material through the nozzle into the mold.

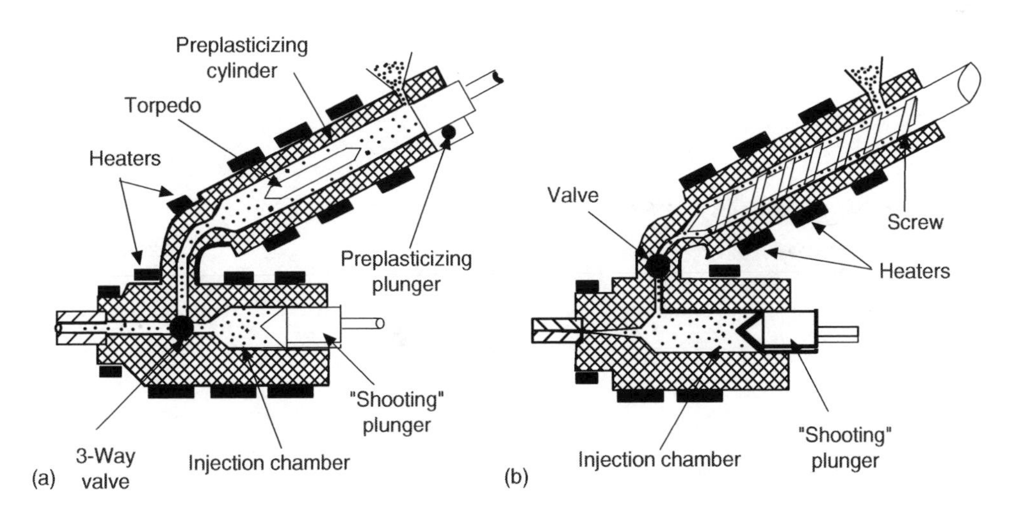

FIGURE 1.7 Schematic drawings of (a) a plunger-type preplasticizer and (b) a screw-type preplasticizer atop a plunger-type injection molding machine. (After *Petrothene: A Processing Guide*, 3rd Ed., 1965. U.S. Industrial Chemicals Co., New York.)

Later, another variation of the two-stage plunger unit appeared, in which the first plunger stage was replaced by a rotation screw in a cylinder (Figure 1.7b). The screw increases the heat transfer at the walls and also does considerable heating by converting mechanical energy into heat. Another advantage of the screw is its mixing and homogenizing action. The screw feeds the melt into the second plunger unit, where the injection ram pushes it forward into the mold.

Although the single-stage plunger units (Figure 1.6) are inherently simple the limited heating rate has caused a decline in popularity: they have been mostly supplanted by the reciprocating screw-type machines. In these machines (Figure 1.8) the plunger and torpedo (or spreader) that are the key components of plunger-type machines are replaced by a rotating screw that moves back and forth like a plunger within the heating cylinder. As the screw rotates, the flights pick up the feed of granular material dropping from the hopper and force it along the heated wall of the barrel, thereby increasing the rate of heat transfer and also generating considerable heat by its mechanical work. The screw, moreover, promotes mixing and homogenization of the plastic material.

As the molten plastic comes off the end of the screw, the screw moves back to permit the melt to accumulate. At the proper time the screw is pushed forward without rotation, acting just like a plunger and forcing the melt through the nozzle into the mold. The size of the charge per shot is regulated by the back travel of the screw. The heating and homogenization of the plastics material are controlled by the screw rotation speed and wall temperatures.

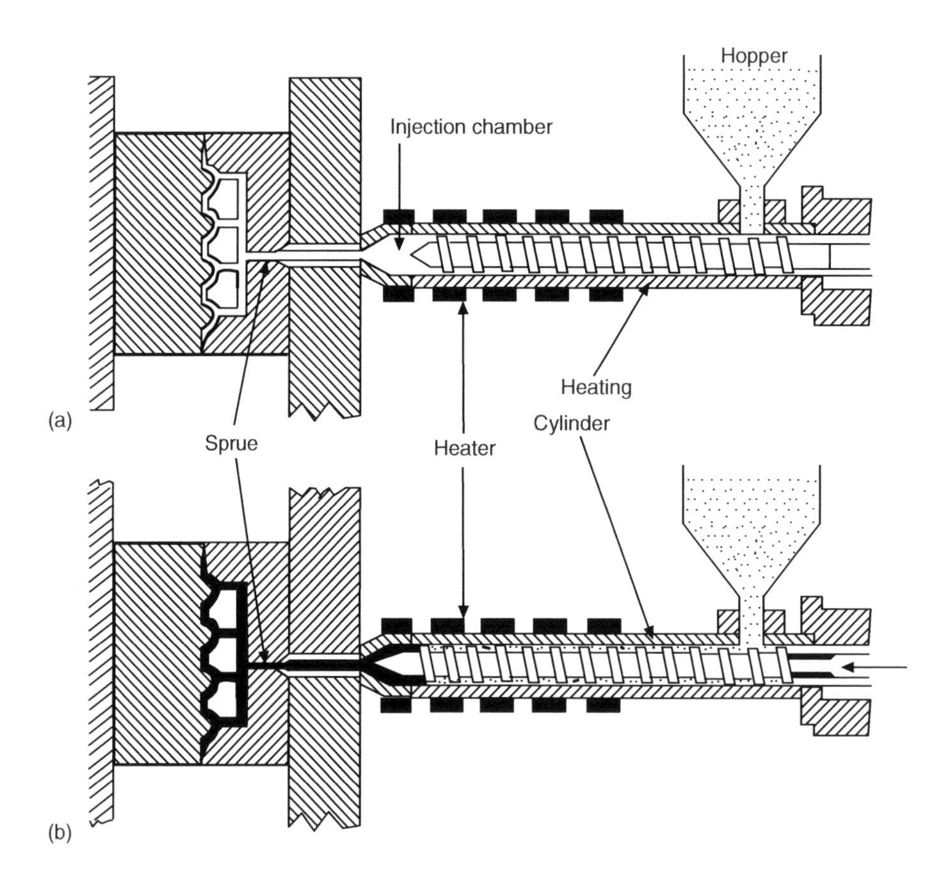

FIGURE 1.8 Cross-section of a typical screw-injection molding machine, showing the screw (a) in the retracted position and (b) in the forward position. (After *Petrothene: A Processing Guide*, 3rd Ed., 1965. U.S. Industrial Chemicals Co., New York.)

1.5.2 Clamping Units

The clamping unit keeps the mold closed while plasticized material is injected into it and opens the mold when the molded article is ejected. The pressure produced by the injection plunger in the cylinder is transmitted through the column of plasticized material and then through the nozzle into the mold. The unlocking force, that is, the force which tends to open the mold, is given by the product of the injection pressure and the projected area of the molding. Obviously, the clamping force must be greater than the unlocking force to hold the mold halves closed during injection.

Several techniques can be used for the clamping unit: (1) hydraulic clamps, in which the hydraulic cylinder operates on the movable parts of the mold to open and close it; (2) toggle or mechanical clamps, in which the hydraulic cylinder operates through a toggle linkage to open and close the mold; and (3) various types of hydraulic mechanical clamps that combine features of (1) and (2).

Clamps are usually built as horizontal units, with injection taking place through the center of the stationary platen, although vertical clamp presses are also available for special jobs.

1.5.3 Molds

The mold is probably the most important element of a molding machine. Although the primary purpose of the mold is to determine the shape of the molded part, it performs several other jobs. It conducts the hot plasticized material from the heating cylinder to the cavity, vents the entrapped air or gas, cools the part until it is rigid, and ejects the part without leaving marks or causing damage. The mold design, construction, the craftsmanship largely determine the quality of the part and its manufacturing cost.

The injection mold is normally described by a variety of criteria, including (1) number of cavities in the mold; (2) material of construction, e.g., steel, stainless steel, hardened steel, beryllium copper, chrome-plated aluminum, and epoxy steel; (3) parting line, e.g., regular, irregular, two-plate mold, and three-plate mold; (4) method of manufacture, e.g., machining, hobbing, casting, pressure casting, electroplating, and spark erosion; (5) runner system, e.g., hot runner and insulated runner; (6) gating type, e.g., edge, restricted (pinpoint), submarine, sprue, ring, diaphragm, tab, flash, fan, and multiple; and (7) method of ejection, e.g., knockout (KO) pins, stripper ring, stripper plate, unscrewing cam, removable insert, hydraulic core pull, and pneumatic core pull.

1.5.3.1 Mold Designs

Molds used for injection molding of thermoplastic resins are usually flash molds, because in injection molding, as in transfer molding, no extra loading space is needed. However, there are many variations of this basic type of mold design.

The design most commonly used for all types of materials is the two plate design (Figure 1.9). The cavities are set in one plate, the plungers in the second plate. The sprue blushing is incorporated in that plate mounted to the stationary half of the mold. With this arrangement it is possible to use a direct center gate that leads either into a single-cavity mold or into a runner system for a multi-cavity mold. The plungers are ejector assembly and, in most cases, the runner system belongs to the moving half of the mold. This is the basic design of an injection mold, though many variations have been developed to meet specific requirements.

A three-plate mold design (Figure 1.10) features a third, movable, plate which contains the cavities, thereby permitting center or offset gating into each cavity for multicavity operation. When the mold is opened, it provides two openings, one for ejection of the molded part and the other for removal of the runner and sprue.

Moldings with inserts or threads or coring that cannot be formed by the normal functioning of the press require installation of separate or loose details or cores in the mold. These loose members are ejected with the molding. They must be separated from the molding and reinstalled in the mold after every cycle. Duplicate sets are therefore used for efficient operation.

Hydraulic or pneumatic cylinders may be mounted on the mold to actuate horizontal coring members. It is possible to mold angular coring, without the need for costly loose details, by adding angular core pins

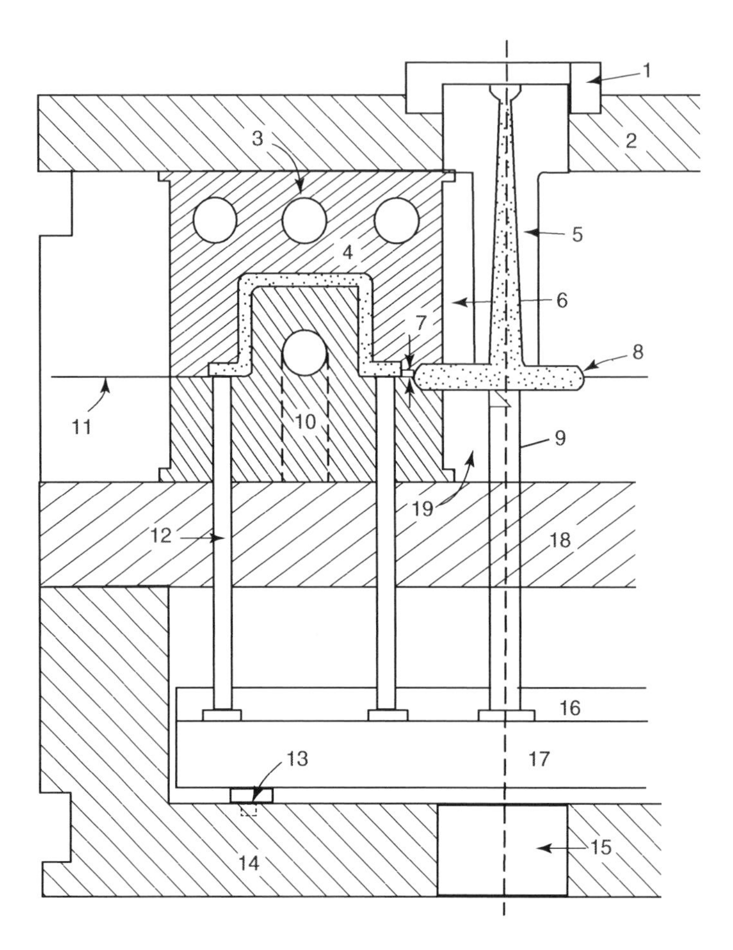

FIGURE 1.9 A two-plate injection-mold design: (1) locating ring; (2) clamping plate; (3) water channels; (4) cavity; (5) sprue bushing; (6) cavity retainer; (7) gate; (8) full round runner; (9) sprue puller pin; (10) plunger; (11) parting line; (12) ejector pin; (13) stop pin; (14) ejector housing; (15) press ejector clearance; (16) pin plate; (17) ejector bar; (18) support plate; (19) plunger retainer.

engaged in sliding mold members. Several methods may be used for unscrewing internal or external threads on molded parts: For high production rates automatic unscrewing may be done at relatively low cost by the use of rack-and-gear mechanism actuated by a double-acting hydraulic long-stroke cylinder. Other methods of unscrewing involve the use of an electric gear-motor drive or friction-mold wipers actuated by double-acting cylinders. Parts with interior undercuts can be made in a mold which has provision for angular movement of the core, the movement being actuated by the ejector bar that frees the metal core from the molding.

1.5.3.2 Number of Mold Cavities

Use of multiple mold cavities permits greater increase in output speeds. However, the greater complexity of the mold also increases significantly the manufacturing cost. Note that in a single-cavity mold the limiting factor is the cooling time of the molding, but with more cavities in the mold the plasticizing capacity of the machine tends to be the limiting factor. Cycle times therefore do not increase prorate with the number of cavities.

There can be no clear-cut answer to the question of optimum number of mold cavities, since it depends on factors such as the complexity of the molding, the size and type of the machine, cycle time,

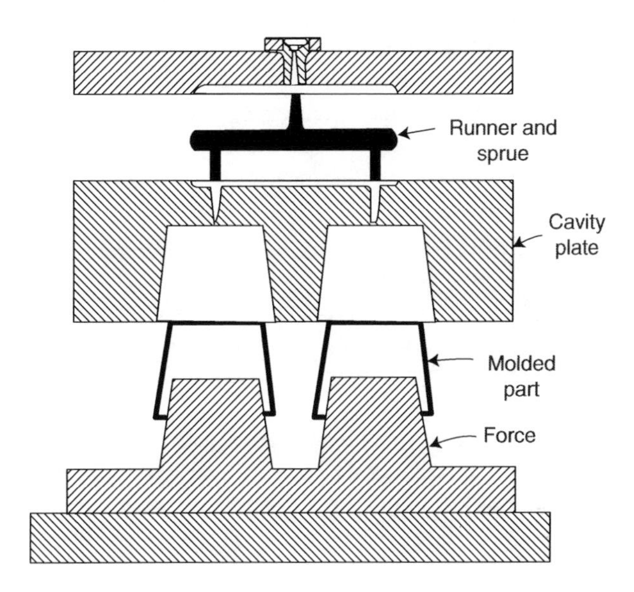

FIGURE 1.10 A diagram of a three-plate injection mold.

and the number of moldings required. If a fairly accurate estimate can be made of the costs and cycle time for molds with each possible number of cavities and a cost of running the machine (with material) is assumed, a break-even quantity of the number of moldings per hour can be calculated and compared with the total production required.

1.5.3.3 Runners

The channels through which the plasticized material enters the gate areas of the mold cavities are called *runners*. Normally, runners are either full round or trapezoidal in cross section. Round cross section offers the least resistance to the flow of material but requires a duplicate machining operation in the mold, since both plates must be cut at the parting line. In three-plate mold designs, however, trapezoidal runners are preferred, since sliding movements are required across the parting-line runner face.

One can see from Figure 1.10 that a three-plate mold operation necessitates removal of the runner and sprue system, which must be reground, and the material reused. It is possible, however, to eliminate the runner system completely by keeping the material in a fluid state. This mold is called a hot-runner mold. The material is kept fluid by the hot-runner manifold, which is heated with electric cartridges.

The advantage of a hot-runner mold is that in a long-running job it is the most economical way of molding—there is no regrinding, with its attendant cost of handling and loss of material, and the mold runs automatically, eliminating variations caused by operators. A hot-runner mold also presents certain difficulties: It takes considerably longer to become operational, and in multicavity molds balancing the gate and the flow and preventing drooling are difficult. These difficulties are partially overcome in an insulated-runner mold, which is a cross between a hot-runner mold and a three-plate mold and has no runner system to regrind. An insulated-runner mold is more difficult to start and operate than a three-plate mold, but it is considerably easier than a hot-runner mold.

1.5.3.4 Gating

The gate provides the connection between the runner and the mold cavity. It must permit enough material to flow into the mold to fill out the cavity. The type of the gate and its size and location in the mold strongly affect the molding process and the quality of the molded part. There are two types of gates: large and restricted. Restricted (pinpointed) gates are usually circular in cross section and for most thermoplastics do not exceed 0.060 in. in diameter. The apparent viscosity of a thermoplastic is a function of shear rate—the viscosity decreases as the shear rate and, hence, the velocity increases. The use of the restricted gate is therefore advantageous, because the velocity of the plastic melt increases as it is forced through the small opening; in addition, some of the kinetic energy is transformed into heat, raising the local temperature of the plastic and thus further reducing its viscosity. The passage through a restricted area also results in higher mixing.

The most common type of gate is the edge gate (Figure 1.11a), where the part is gated either as a restricted or larger gate at some point on the edge. The edge gate is easy to construct and often is the only practical way of gating. It can be fanned out for large parts or when there is a special reason. Then it is called a fan gate (Figure 1.11f). When it is required to orient the flow pattern in one direction, a flash gate (Figure 1.11c) may be used. It involves extending the fan gate over the full length of the part but keeping it very thin.

The most common gate for single-cavity molds is the sprue gate (Figure 1.11d). It feeds directly from the nozzle of the machine into the molded part. The pressure loss is therefore a minimum. But the sprue gate has the disadvantages of the lack of a cold slug, the high stress concentration around the gate area, and the need for gate removal. A diaphragm gate (Figure 1.11e) has, in addition to the sprue, a circular area leading from the sprue to the piece. This type of gate is suitable for gating hollow tubes. The diaphragm eliminates stress concentration around the gate because the whole area is removed, but the cleaning of this gate is more difficult than for a sprue gate. Ring gates (Figure 1.11f) accomplish the same purpose as gating internally in a hollow tube, but from the outside.

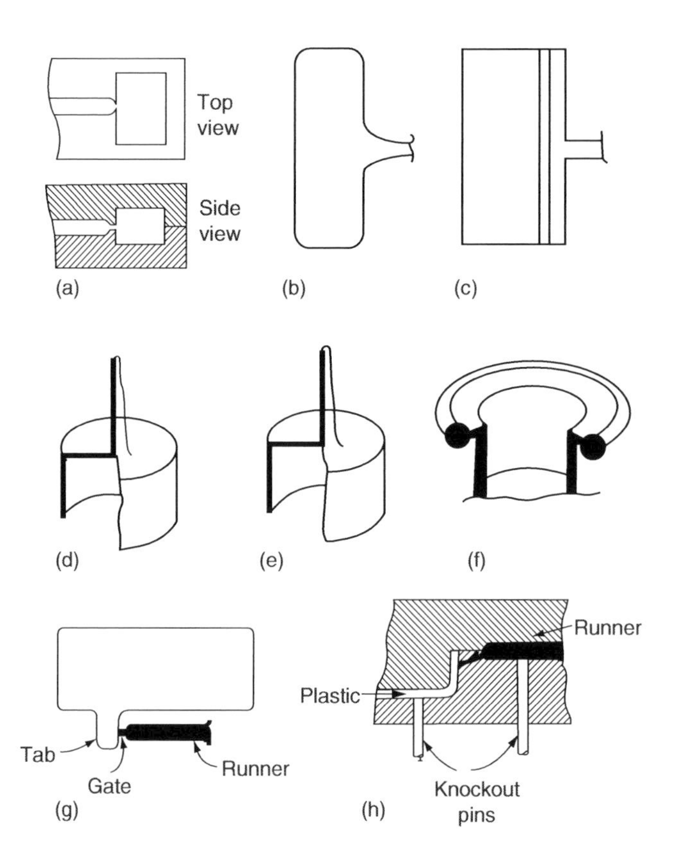

FIGURE 1.11 Gating design: (a) edge; (b) fan; (c) flash; (d) sprue; (e) diaphragm; (f) ring; (g) tab; (h) submarine.

When the gate leads directly into the part, there may be surface imperfection due to jetting. This may be overcome by extending a tab from the part into which the gate is cut. This procedure is called tab gating (Figure 1.14g). The tab has to be removed as a secondary operation.

A submarine gate (Figure 1.11h) is one that goes through the steel of the cavity. It is very often used in automatic molds.

1.5.3.5 Venting

When the melted plastic fills the mold, it displaces the air. The displaced air must be removed quickly, or it may ignite the plastic and cause a characteristic burn, or it may restrict the flow of the melt into the mold cavity, resulting in incomplete filling. For venting the air from the cavity, slots can be milled, usually opposite the gate. The slots usually range from 0.001 to 0.002 in. deep and from 3/8 to 1 in. wide. Additional venting is provided by the clearance between KO pins and their holes. Note that the gate location is directly related to the consideration of proper venting.

1.5.3.6 Parting Line

If one were inside a closed mold and looking outside, the mating junction of the mold cavities would appear as a line. It also appears as a line on the molded piece and is called the parting line. A piece may have several parting lines. The selection of the parting line in mold design is influenced by the type of mold, number of cavities, shape of the piece, tapers, method of ejection, method of fabrication, venting, wall thickness, location and type of gating, inserts, postmolding operations, and aesthetic consideration.

1.5.3.7 Cooling

The mold for thermoplastics receives the molten plastic in its cavity and cools it to solidity to the point of ejection. The mold is provided with cooling channels. The mold temperature is controlled by regulating the temperature of the cooling fluid and its rate of flow through the channels. Proper cooling or coolant circulation is essential for uniform repetitive mold cycling.

The functioning of the mold and the quality of the molded part depend largely on the location of the cooling channel. Since the rate of heat transfer is reduced drastically by the interface of two metal pieces, no matter how well they fit, cooling channels should be located in cavities and cores themselves rather than only in the supporting plates. The cooling channels should be spaced evenly to prevent uneven temperatures on the mold surface. They should be as close to the plastic surface as possible, taking into account the strength of the mold material. The channels are connected to permit a uniform flow of the cooling or heating medium, and they are thermostatically controlled to maintain a given temperature.

Another important factor in mold temperature control is the material the mold is made from. Beryllium copper has a high thermal conductivity, about twice that of steel and four times that of stainless steel. A beryllium copper cavity should thus cool about four times as fast as a stainless steel one. A mold made of beryllium copper would therefore run significantly faster than one of stainless steel.

1.5.3.8 Ejection

Once the molded part has cooled sufficiently in the cavity, it has to be ejected. This is done mechanically by KO pins, KO sleeves, stripper plates, stripper rings or compressed air, used either singly or in combination. The most frequent problem in new molds is with ejection. Because there is no mathematical way of predicting the amount of ejection force needed, it is entirely a matter of experience.

Since ejection involves overcoming the forces of adhesion between the mold and the plastic, the area provided for the knockout (KO) is an important factor. If the area is too small, the KO force will be concentrated, resulting in severe stresses on the part. As a result, the part may fail immediately or in later service. In materials such as ABS and high-impact polystyrene, the severe stresses can also discolor the plastic.

Sticking in a mold makes ejection difficult. Sticking is often related to the elasticity of steel and is called packing. When injection pressure is applied to the molten plastic and force it into the mold, the steel deforms; when the pressure is relieved, the steel retracts, acting as a clamp on the plastic. Packing is often eliminated by reducing the injection pressure and/or the injection forward time. Packing is a common problem in multicavity molds and is caused by unequal filling. Thus, if a cavity seals off without filling, the material intended for the cavity is forced into other cavities, causing overfilling.

1.5.3.9 Standard Mold Bases

Standardization of mold bases for injection molding, which was unknown prior to 1940, was an important factor in the development of efficient mold making. Standard mold bases were pioneered by the D-M-E Co., Michigan, to provide the mold maker with a mold base at lower cost and with much higher quality than if the base were manufactured by the mold maker. Replacement parts, such as locating ring and sprue bushings, loader pins and bushings, KO pins and push-back pins of high quality are also available to the molder. Since these parts are common for many molds, they can be stocked by the molder in the plant and thus down time is minimized. An exploded view of the components of a standard injection-mold base assembly is shown in Figure 1.12.

1.5.4 Structural Foam Injection Molding

Structural foam injection molding produces parts consisting of solid external skin surfaces surrounding an inner cellular (or foam) core, as shown in Figure 1.13. Large, thick structural foam parts can be produced by this process with both low and high pressure and using either nitrogen gas or chemical blowing agents (see "Foaming Process").

1.5.5 Co-Injection (Sandwich) Molding

Co-injection molding is used to produce parts that have a laminated structure with the core material embedded between the layers of the skin material. As shown in Figure 1.14, the process involves sequential injection of two different but compatible polymer melts into a cavity where the materials

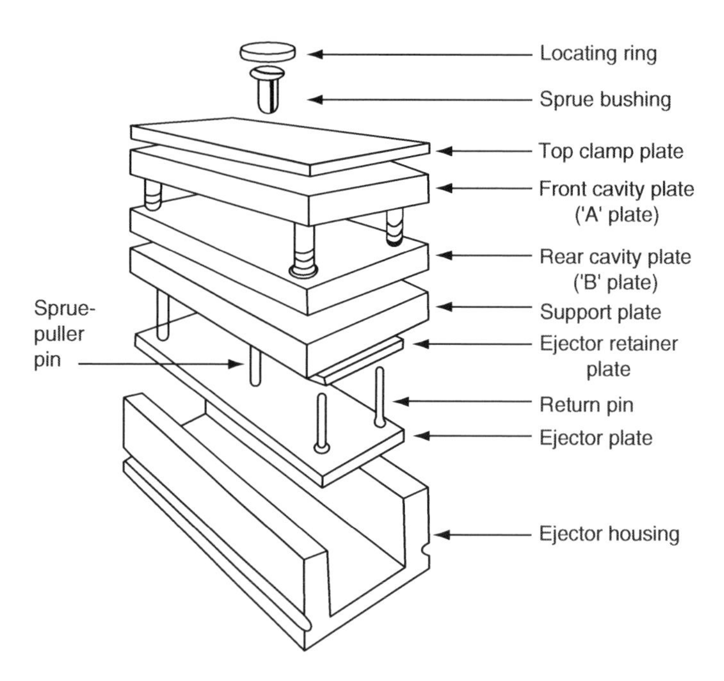

FIGURE 1.12 Exploded view of a standard mold base showing component parts.

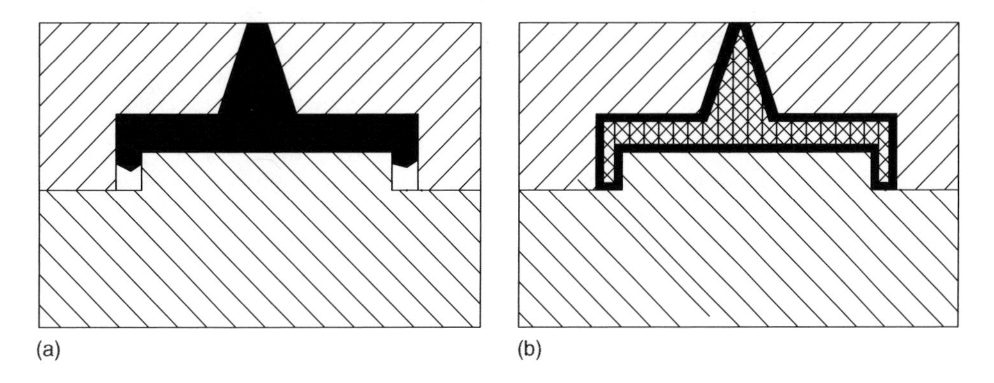

FIGURE 1.13 Structural foam injection molding. (a) During injection under high pressure there is very little foaming. (b) After injection, pressure drops and foaming occurs at hot core.

laminate and solidify. A short shot of skin polymer melt is first injected into the mold (Figure 1.14a), followed by core polymer melt which is injected until the mold cavity is nearly filled (Figure 1.14b); the skin polymer is then injected again to purge the core polymer away from the spruce (Figure 1.14c). The process offers the inherent flexibility of using the optimal properties of each material or modifying the properties of each material or those of the molded part.

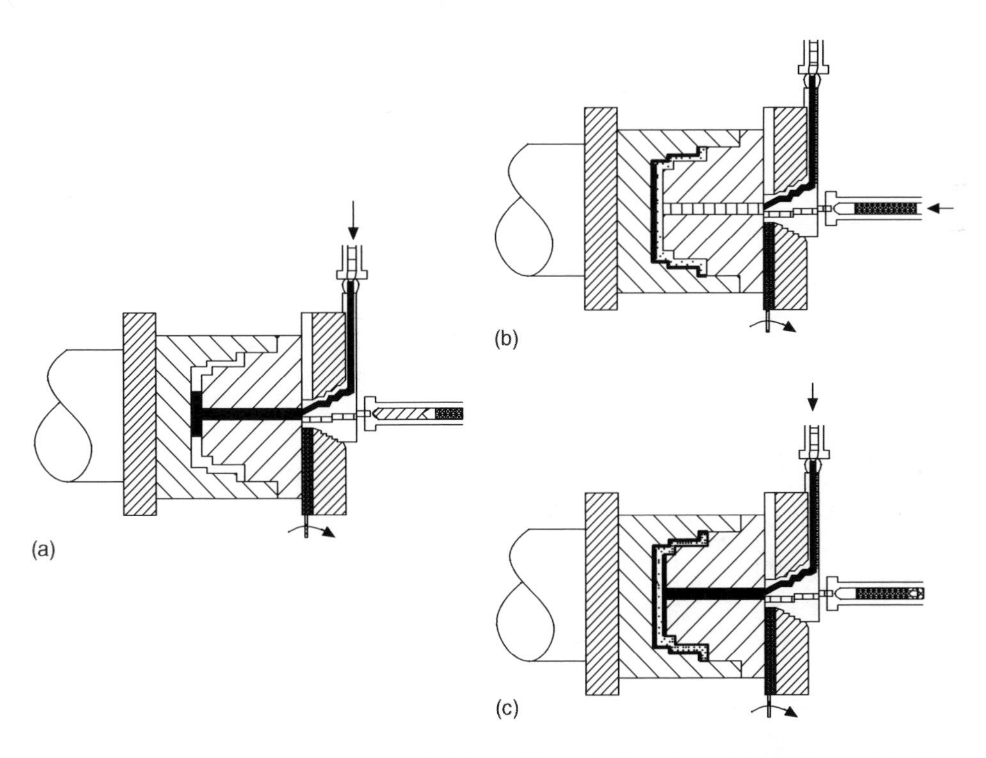

FIGURE 1.14 Three stages of co-injection (sandwich) molding. (a) Short shot of skin polymer melt (shown in black) is injected into the mold; (b) injection of core polymer melt until cavity is nearly filled; and (c) skin polymer melt is injected again, pushing the core polymer away from the sprue.

1.5.6 Gas-Assisted Injection Molding

The gas-assisted injection molding process begins with a partial or full injection of polymer melt into the mold cavity. Compressed gas is then injected into the core of the polymer melt to help fill and pack the mold, as shown in Figure 1.15 for the Asahi Gas Injection Molding process. This process is thus capable of producing hollow rigid parts, free of sink marks. The hollowing out of thick sections of moldings results in reduction in part weight and saving of resin material.

Other advantages include shorter cooling cycles, reduced clamp force tonnage and part consolidation. The process allows high precision molding with greater dimensional stability by eliminating uneven mold shrinkage and makes it possible to mold complicated shapes in single form, thus reducing product assembly work and simplifying mold design.

The formation of thick walled sections of a molding can be easily achieved by introducing gas in the desired locations. The gas channels thus formed also effectively support the flow of resin, allowing the molding pressure to be greatly reduced, which in turn reduces internal stresses, allows uniform mold shrinkage, and reduces sink marks and warpage.

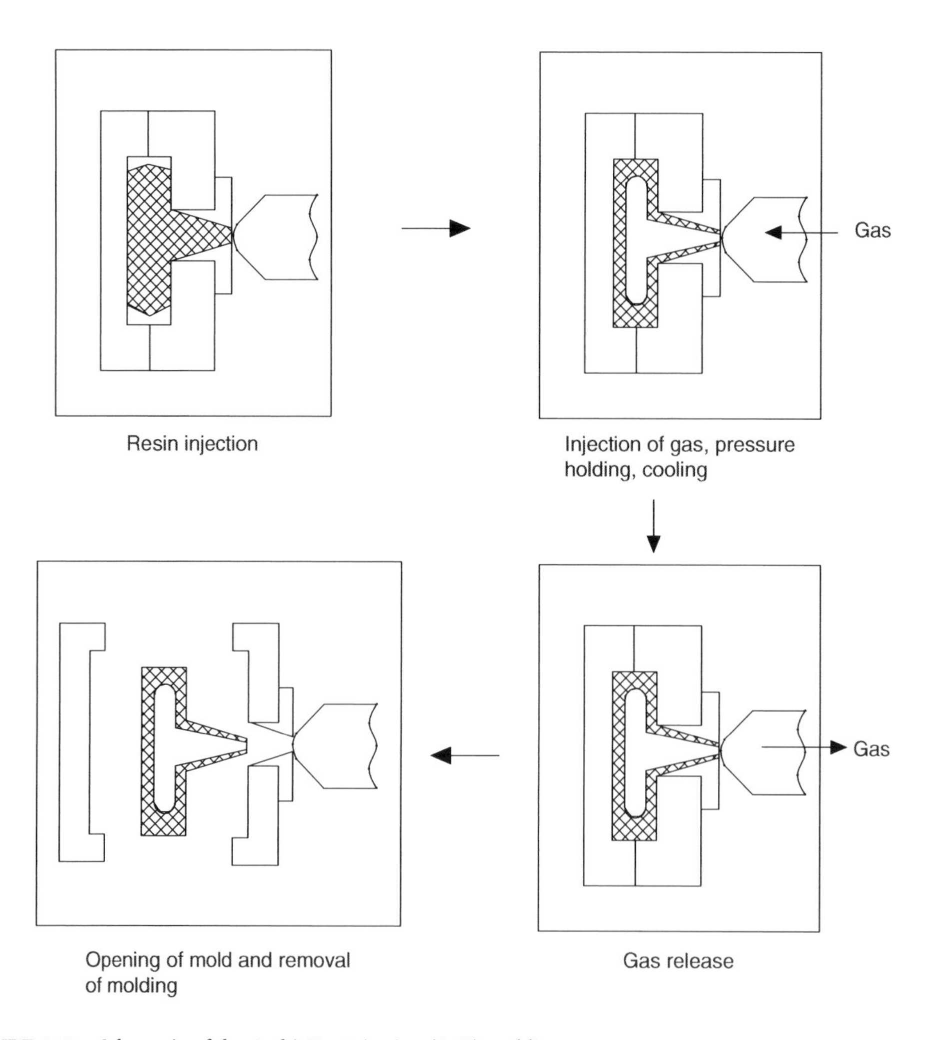

FIGURE 1.15 Schematic of the Asahi Gas Injection (AGI) molding process.

1.6 Injection Molding of Thermosetting Resins

1.6.1 Screw-Injection Molding of Thermosetting Resins

The machines used earlier were basically plunger-type machines [6,10–12]. But in the late 1960s' shortly after the development of screw-transfer machines, the concept of screw-injection molding of thermosets, also known as direct screw transfer (or DST), was introduced. The potential of this technique for low-cost, high-volume production of molded thermoset parts was quickly recognized, and today screw-injection machines are available in all clamp tonnages up to 1,200 tons and shot sizes up to 10 lb. Coupled with this, there has been a new series of thermosetting molding materials developed specifically for injection molding. These materials have long life at moderate temperature (approximately 200°F), which permits plastication in screw barrel, and react (cure) very rapidly when the temperature is raised to 350°F–400°F (177°–204°C), resulting in reduced cycle time.

A typical arrangement for a direct screw-transfer injection-molding machine for thermosets is shown in Figure 1.16. The machine has two sections mounted on a common base. One section constitutes the plasticizing and injection unit, which includes the feed hopper, the heated barrel that encloses the screw, the hydraulic cylinder which pushes the screw forward to inject the plasticized material into the mold, and a motor to rotate the screw. The other section clamps and holds the mold halves together under pressure during the injection of the hot plastic melt into the mold.

The thermosetting material (in granular or pellet form) suitable for injection molding is fed from the hopper into the barrel and is then moved forward by the rotation of the screw. During its passage, the material receives conductive heat from the wall of the heated barrel and frictional heat from the rotation of the screw. For thermosetting materials, the screw used is a zero-compression-ratio screw—i.e., the depths of flights of the screw at the feed-zone end and at the nozzle end are the same. By comparison, the screws used in thermoplastic molding machines have compression ratios such that the depth of flight at the feed end is one and one-half to five times that at the nozzle end. This difference in screw configuration is a major difference between thermoplastic- and thermosetting-molding machines.

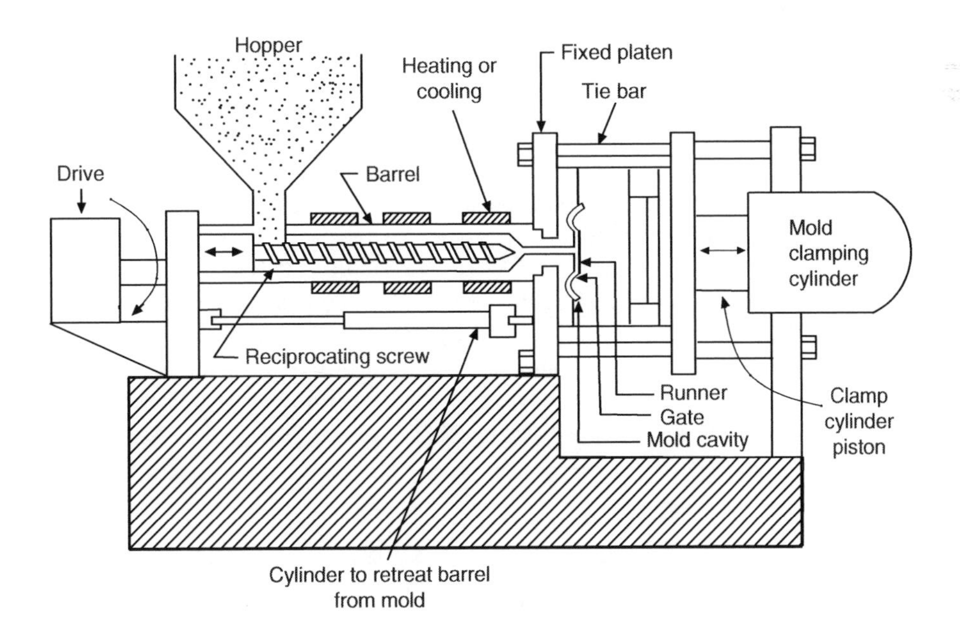

FIGURE 1.16 Schematic of a direct screw-transfer molding machine for thermosets. (After Frados, J. ed. 1976. *Plastics Engineering Handbook, 4th Ed.*, Van Nostrand Reinhold, New York.)

As the material moves forward in the barrel due to rotation of the screw, it changes in consistency from a solid to semifluid, and as it starts accumulating at the nozzle end, it exerts a backward pressure on the screw. This back pressure is thus used as a processing variable. The screw stops turning when the required amount of material—the charge—has accumulated at the nozzle end of the barrel, as sensed by a limit switch. (The charge is the exact volume of material required to fill the sprue, runners, and cavities of the mold.) The screw is then moved forward like a plunger by hydraulic pressure (up to 20,000 psi) to force the hot plastic melt through the sprue of the mold and into the runner system, gates, and mold cavities. The nozzle temperature is controlled to maintain a proper balance between a hot mold (350°F–400°F), and a relatively cool barrel (150°F–200°F).

Molded-in inserts are commonly used with thermosetting materials. However, since the screw-injection process is automatic, it is desirable to use postassembled inserts rather than molded-in inserts because molded-in inserts require that the mold be held open each cycle to place the inserts. A delay in the manual placement disrupts an automatic cyclic operation, affecting both the production rate and the product quality.

Tolerances of parts made by injection molding of thermosetting materials are comparable to those produced by the compression and transfer methods described, earlier. Tolerances achieved are as low as ± 0.001 in./in., although ordinarily tolerance of ± 0.003–0.005 in./in. are economically practical.

Thermosetting materials used in screw-injection molding are modified from conventional thermosetting compounds. These modifications are necessary to provide the working time-temperature relationship required for screw plasticating. The most commonly used injection-molding thermosetting materials are the phenolics. Other thermosetting materials often molded by the screw-injection process include melamine, urea, polyester, alkyd, and dially phthalate (DAP).

Since the mid-1970s the injection molding of glass-fiber-reinforced thermosetting polyesters gained increasing importance as better materials (e.g., low shrinkage resins, palletized forms of polyester/glass, etc.), equipment, and tooling became available. Injection-molded reinforced thermoset plastics have thus made inroads in such markets as switch housings, fuse blocks, distributor caps, power-tool housings, office machines, etc. Bulk molding compounds (BMC), which are puttylike FRP (fibrous glass-reinforced plastic) materials, are injection molded to make substitutes of various metal die castings.

For injection molding, FRP should have some specific characteristics. For example, it must flow easily at lower-than-mold temperatures without curing and without separating into resin, glass, and filler components, and it should cure rapidly when in place at mold temperature. A traditional FRP material shrinks about 0.003 in./in. during molding, but low-shrink FRP materials used for injection molding shrink as little as 0.000–0.0005 in./in. Combined with proper tooling, these materials thus permit production of pieces with dimensional tolerances of ± 0.0005 in./in.

Proper design of parts for injection molding requires an understanding of the flow characteristics of material within the mold. In this respect, injection-molded parts of thermosets are more like transfer-molded parts than to compression-molded parts. Wall-section uniformity is an important consideration in part design. Cross sections should be as uniform as possible, within the dictates of part requirements, since molding cycles, and therefore costs, depend on the cure time of the thickest section. (For thermoplastics, however, it is the cooling time that is critical). A rule of thumb for estimating cycle times for a 1/4-in. wall section is 30 sec for injection-molded thermosets (compared to 45 sec for thermoplastics). As a guideline for part design, a good working average for wall thickness is 1/8–3/16 in., with a minimum of 1/16 in.

1.7 Extrusion

The extrusion process is basically designed to continuously convert a soft material into a particular form [13–15]. An oversimplified analogy may be a house-hold meat grinder. However, unlike the extrudate from a meat grinder, plastic extrudates generally approach truly continuous formation. Like the usual meat grinder, the extruder (Figure 1.17) is essentially a screw conveyor. It carries the cold plastic material

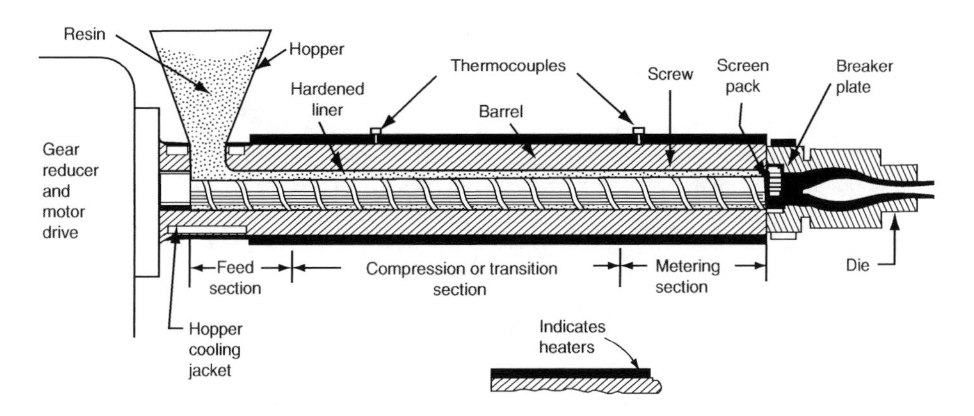

FIGURE 1.17 Scheme for a typical single-screw extruder showing extruding pipe.

(in granular or powdered form) forward by the action of the screw, squeezes it, and, with heat from external heaters and the friction of viscous flow, changes it to a molten stream. As it does this, it develops pressure on the material, which is highest right before the molten plastic enters the die. The screen pack, consisting of a number of fine or coarse mesh gauzes supported on a breaker plate and placed between the screw and the die, filter out dirt and unfused polymer lumps. The pressure on the molten plastic forces it through an adapter and into the die, which dictates the shape of the fine extrudate. A die with a round opening as shown in Figure 1.17, produces pipe; a square die opening produces a square profile, etc. Other continuous shapes, such as the film, sheet, rods, tubing, and filaments, can be produced with appropriate dies. Extruders are also used to apply insulation and jacketing to wire and cable and to coat substrates such as paper, cloth, and foil.

When thermoplastic polymers are extruded, it is necessary to cool the extrudate below T_m or T_g to impart dimensional stability. This cooling can often be done simply by running the product through a tank of water, by spraying cold water, or, even more simply, by air cooling. When rubber is extruded, dimensional stability results from cross-linking (vulcanization). Interestingly, rubber extrusion for wire coating was the first application of the screw extruder in polymer processing.

Extruders have several other applications in polymer processing: in the blowmolding process they are used to make hollow objects such as bottles; in the blow-film process they are used for making wide films; they are also used for compounding plastics (i.e., adding various ingredients to a resin mix) and for converting plastics into the pellet shape commonly used in processing. In this last operation specialized equipment, such as the die plate-cutter assembly, is installed in place of the die, and an extrusion-type screw is used to provide plasticated melt for various injection-molding processes.

1.7.1 Extruder Capacity

Standard sizes of single-screw extruders are $1\frac{1}{2}$, 2, $2\frac{1}{2}$, $3\frac{1}{4}$, $3\frac{1}{2}$, $4\frac{1}{2}$, 6, and 8 in., which denote the inside diameter (ID) of the barrel. As a rough guide, extruder capacity Q_e, in pounds per hour, can be calculated from the barrel diameter D_b, in inches, by the empirical relation [15]

$$Q_e = 16 D_b^{2.2} \tag{1.1}$$

Another estimate of extruder capacity can be made by realizing that most of the energy needed to melt the thermoplastic stems from the mechanical work, whereas the barrel heaters serve mainly to insulate the material. If we allow an efficiency from drive to screw of about 80%, the capacity Q_e (lb/h) can be approximately related to the power supplied H_p (horsepower), the heat capacity of the material C_p [Btu/lb °F], and the temperature rise from feed to extrudate ΔT (°F) by

$$Q_e = 1.9 \times 10^3 H_p / C_p \Delta T \tag{1.2}$$

Equation 1.2 is obviously not exact since the heat of melting and other thermal effects have been ignored. Equation 1.2 coupled with Equation 1.1 enables one to obtain an estimate of ΔT. Thus, for processing of poly(methyl methacrylate), for which C_p is about 0.6 Btu/lb °F, in a 2-in. extruder run by a 10-hp motor, Equation 1.1 gives $Q_e = 74$ lb/h, and Equation 1.2 indicates that $\Delta T \cong 430$°F. In practice, a ΔT of 350°F is usually adequate for this polymer.

1.7.2 Extruder Design and Operation

The most important component of any extruder is the screw. It is often impossible to extrude satisfactorily one material by using a screw designed for another material. Therefore screw designs vary with each material.

1.7.2.1 Typical Screw Construction

The screw consists of a steel cylinder with a helical channel cut into it (Figure 1.18). The helical ridge formed by the machining of the channel is called the flight, and the distance between the flights is called the lead. The lead is usually constant for the length of the screw in single-screw machines. The helix angle is called pitch. Helix angles of the screw are usually chosen to optimize the feeding characteristics. An angle of 17.5° is typical, though it can be varied between 12 and 20°. The screw outside diameter is generally just a few thousandths of an inch less than the ID of the barrel. The minimal clearance between screw and barrel ID prevents excessive buildup of resin on the inside barrel wall and thus maximizes heat transfer.

The screw may be solid or cored. Coring is used for steam heating or, more often, for water cooling. Coring can be for the entire length of the screw or for a portion of it, depending on the particular application. Full length coring of the screw is used where large amounts of heat are to be removed. The screw is cored only in the initial portions at the hopper end when the objective is to keep the feed zone cooler for resins which tend to soften easily. Screws are often fabricated from 4140 alloy steel, but other materials are also used. The screw flights are usually hardened by flame-hardening techniques or inset with a wear resistant alloy (e.g., Stellite 6).

1.7.2.2 Screw Zones

Screws are characterized by their length-diameter ratio (commonly written as L/D ratios). L/D ratios most commonly used for single-screw extruders range from 15:1 to 30:1. Ratios of 20:1 and 24:1 are common for thermoplastics, whereas lower values are used for rubbers. A long barrel gives a more homogeneous extrudate, which is particularly desirable when pigmented materials are handled. Screws are also characterized by their compression ratios—the ratio of the flight depth of the screw at the hopper end to the flight depth at the die end. Compression ratios of single-screw extruders are usually 2:1–5:1.

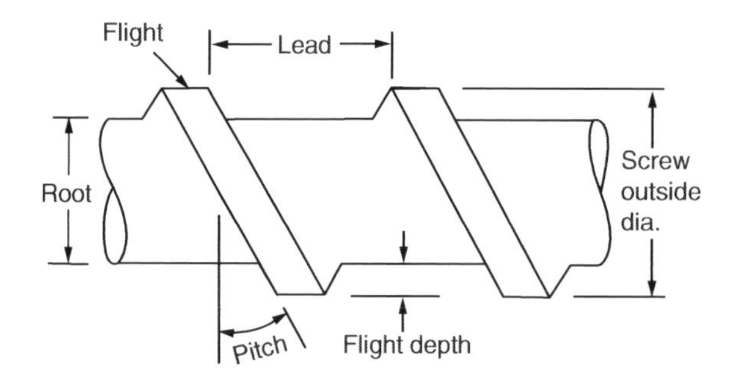

FIGURE 1.18 Detail of screw.

The screw is usually divided into three sections, namely, feed, compression, and metering (Figure 1.17). One of the basic parameters in screw design involves the ratio of lengths between the feed, compression (or transition), and metering sections of the screw. Each section has its own special rate. The feed section picks up the powder, pellets, or beads from under the hopper mouth and conveys them forward in the solid state to the compression section. The feed section is deep flighted so that it functions in supplying enough material to prevent starving the forward sections.

The gradually diminishing flight depth in the compression section causes volume compression of the melting granules. The volume compression results in the trapped air being forced back through the feed section instead of being carried forward with the resin, thus ensuring an extrudate free from porosity. Another consequence of volume compression is an increase in the shearing action on the melt, which is caused by the relative motion of the screw surfaces with respect to the barrel wall. The increased shearing action produces good mixing and generates frictional heat, which increases fluidity of the melt and leads to a more uniform temperature distribution in the molten extrudate. The resin should be fully melted into a reasonably uniform melt by the time it enters the final section of the screw, known as the metering section. The function of the metering section is to force the molten polymer through the die at a steady rate and iron out pulsations. For many screw designs the compression ratio is 3–5; i.e., the flight depth in the metering section is one-third to one-fifth that in the feed section.

1.7.2.3 Motor Drive

The motor employed for driving the screw of an extruder should be of more than adequate power required for its normal needs. Variable screw speeds are considered essential. Either variable-speed motors or constant-speed motors with variable-speed equipment, such as hydraulic systems, step-change gear boxes, and infinitely variable-speed gear boxes may be used. Thrust bearings of robust construction are essential because of the very high back pressure generated in an extruder and the trend towards higher screw speeds. Overload protection in the form of an automatic cut-out should be fitted.

1.7.2.4 Heating

Heat to melt the polymer granules is supplied by external heaters or by frictional heat generated by the compression and shearing action of the screw on the polymer. Frictional heat is considerable, and in modern high-speed screw extruders it supplies most of the heat required for steady running. External heaters serve only to insulate the material and to prevent the machine from stalling at the start of the run when the material is cold. The external heater may be an oil, steam, or electrical type. Electrical heating is most popular because it is compact, clean, and accurately controlled. Induction heating is also used because it gives quicker heating with less variation and facilitates efficient cooling.

The barrel is usually divided into three or four heating zones; the temperature is lowest at the feed end and highest at the die end. The temperature of each zone is controlled by carefully balancing heating and cooling. Cooling is done automatically by either air or water. (The screw is also cored for heating and cooling.) The screw is cooled where the maximum amount of compounding is required, because this improves the quality of the extrudate.

1.7.2.5 Screw Design

The screw we have described is a simple continuous-flight screw with constant pitch. The more sophisticated screw designs include flow disrupters or mixing sections (Figure 1.19). These mixer screws have mixing sections which are designed as mechanical means to break up and rearrange the laminar flow of the melt within the flight channel, which results in more thorough melt mixing and more uniform heat distribution in the metering section of the screw.

Mixer screws have also been used to mix dissimilar materials (e.g., resin and additives or simply dissimilar resins) and to improve extrudate uniformity at higher screw speeds (> 100 rpm). A few typical mixing section designs are shown in Figure 1.20. The fluted-mixing-section-barrier-type design (Figure 1.20a) has proved to be especially applicable for extrusion of polyolefins. For some mixing problems, such as pigment mixing during extrusion, it is convenient to use rings (Figure 1.20b)

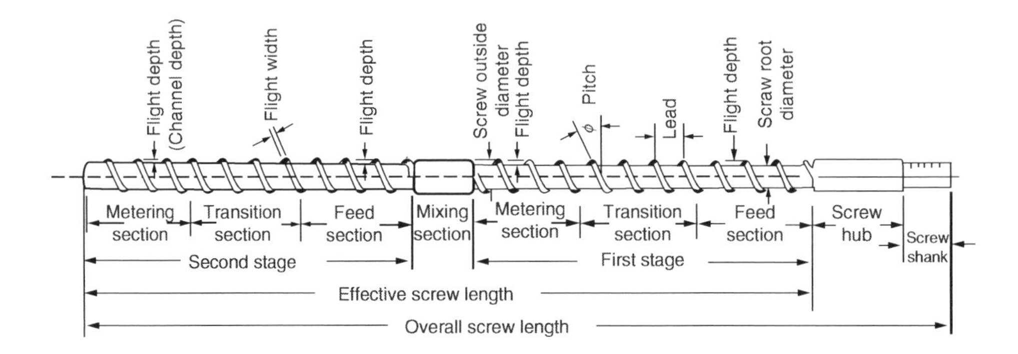

FIGURE 1.19 Single-flight, two-stage extrusion screw with mixing section.

or mixing pins (Figure 1.20c) and sometimes parallel interrupted mixing flights having wide pitch angles (Figure 1.20d).

A later development in extruder design has been the use of venting or degassing zones to remove any volatile constituents from the melt before it is extruded through the die. This can be achieved by placing an obstruction in the barrel (the reverse flights in Figure 1.21) and by using a valved bypass section to step down the pressure developed in the first stage to atmospheric pressure for venting. In effect, two screws are used in series and separated by the degassing or venting zone. Degassing may also be achieved by having a deeper thread in the screw in the degassing section than in the final section of the first screw, so the polymer melt suddenly finds itself in an increased volume and hence is at a lower pressure. The volatile vapors released from the melt are vented through a hold in the top of the extruder barrel or through a hollow core of the screw by way of a hole drilled in the trailing edge of one of the flights in the degassing zone. A vacuum is sometimes applied to assist in the extraction of the vapor. Design and operation must be suitably controlled to minimize plugging of the vent (which, as noted above, is basically an open area) or the possibility of the melt escaping from this area.

Many variations are possible in screw design to accommodate a wide range of polymers and applications. So many parameters are involved, including such variables as screw geometry, materials characteristics, operating conditions, etc., that the industry now uses computerized screw design, which permits analysis of the variables by using mathematical models to derive optimum design of a screw for a given application.

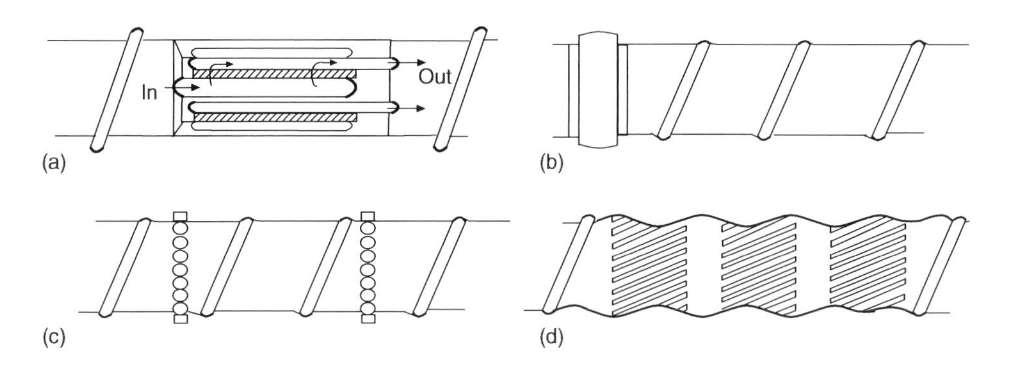

FIGURE 1.20 Mixing section designs: (a) fluted-mixing-section-barrier type; (b) ring-barrier type; (c) mixing pins; (d) parallel interrupted mixing flights.

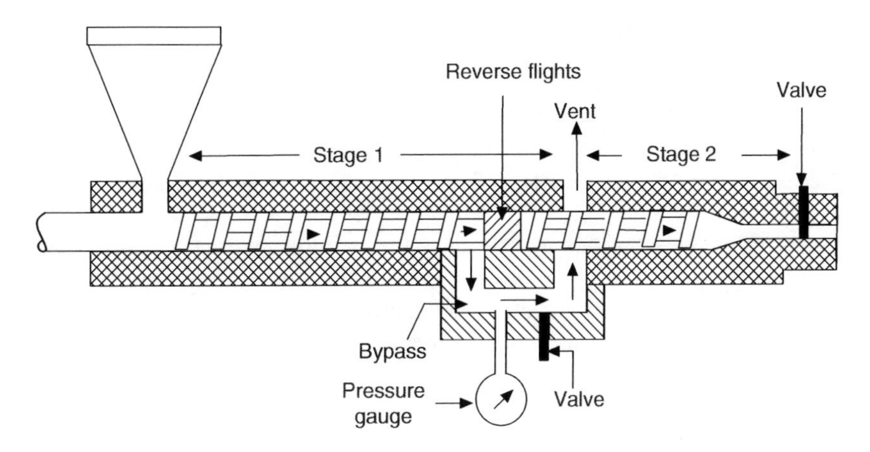

FIGURE 1.21 A two-stage vented extruder with a valved bypass. (After Van Ness, R.T., De Hoff, G.R., and Bonner, R.M. 1968. *Mod. Plastics*, 45, 14A, Encycl. Issue, 672.)

Various screw designs have been recommended by the industry for extrusion of different plastics. For polyethylene, for example, the screw should be long with an *L*/*D* of at least 16:1 or 30:1 to provide a large area for heat transfer and plastication. A constant-pitch, decreasing-channel-depth, metering-type polyethylene screw or constant-pitch, constant-channel-depth, metering-type nylon screw with a compression ratio between 3–1 and 4–1 (Figure 1.22) is recommended for polyethylene extrusion, the former being preferable for film extension and extrusion coating. Nylon-6, 6 melts at approximately 260°C (500°F). Therefore, an extruder with an *L*/*D* of at least 16:1 is necessary. A screw with a compression ratio of 4:1 is recommended.

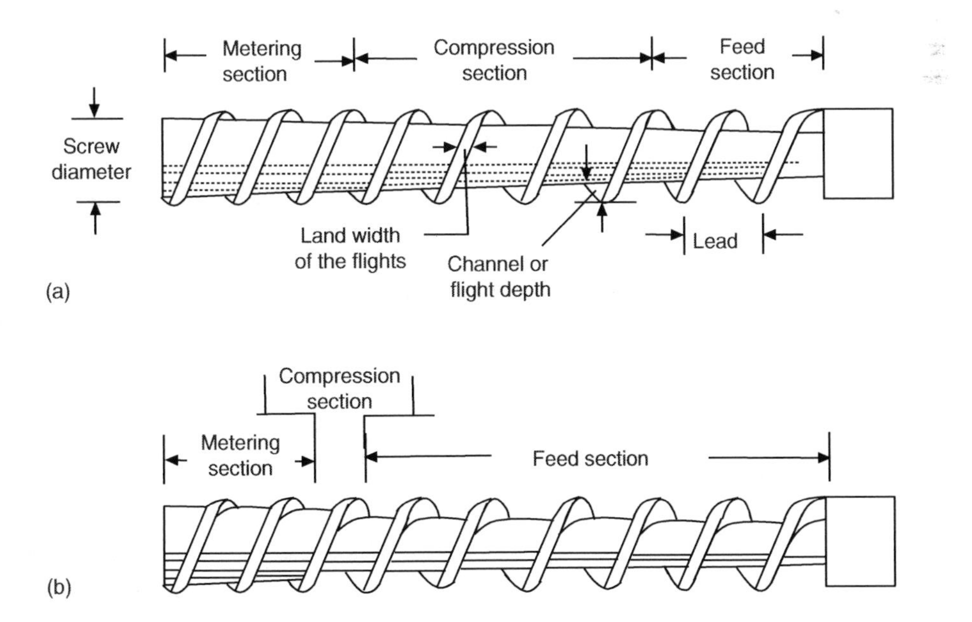

FIGURE 1.22 (a) Constant pitch, decreasing channel depth, metering-type polyethylene screw. (b) Constant pitch, constant-channel-depth, metering-type nylon screw (not to scale).

1.7.3 Multiple-Screw Extruders

Multiple-screw extruders (that is, extruders with more than a single screw) were developed largely as a compounding device for uniformly blending plasticizers, fillers, pigments, stabilizers, etc., into the polymer. Subsequently, the multiple-screw extruders also found use in the processing of plastics.

Multiple-screw extruders differ significantly from single-screw extruders in mode of operation. In a single-screw machine, friction between the resin and the rotating screw, makes the resin rotate with the screw, and the friction between the rotating resin and the barrel pushes the material forward, and this also generates heat. Increasing the screw speed and/or screw diameter to achieve a higher output rate in a single-screw extruder will therefore result in a higher buildup of frictional heat and higher temperatures. In contrast, in twin-screw extruders with intermeshing screws the relative motion of the flight of one screw inside the channel of the other pushes the material forward almost as if the machine were a positive-displacement gear pump which conveys the material with very low friction.

In two-screw extruders, heat is therefore controlled independently from an outside source and is not influenced by screw speed. This fact become especially important when processing a heat-sensitive plastic like poly(vinyl chloride) (PVC). Multiple-screw extruders are therefore gaining wide acceptance for processing vinyls, although they are more expensive than single-screw machines. For the same reason, multiple-screw extruders have found a major use in the production of high-quality rigid PVC pipe of large diameter.

Several types of multiple-screw machines are available, including intermeshing corotating screws (in which the screws rotate in the same direction, and the flight of one screw moves inside the channel of the other), intermeshing counterrotating screws (in which the screws rotate in opposite directions), and nonintermeshing counterrotating screws.

Multiple-screw extruders can involve either two screws (twin-screw design) or four screws. A typical four-screw extruder is a two-stage machine, in which a twin-screw plasticating section feeds into a twin-screw discharge section located directly below it. The multiple screws are generally sized on output rates (lb/h) rather than on *L/D* ratios or barrel diameters.

1.7.4 Blown-Film Extrusion

The blown-film technique is widely used in the manufacture of polyethylene and other plastic films [14,15]. A typical setup is shown in Figure 1.23. In this case the molten polymer from the extruder head enters the die, where it flows round a mandrel and emerges through a ring-shaped opening in the form of a tube. The tube is expanded into a bubble of the required diameter by the pressure of internal air admitted through the center of the mandrel. The air contained in the bubble cannot escape because it is sealed by the die at one end and by the nip (or pinch) rolls at the other, so it acts like a permanent shaping mandrel once it has been injected. An even pressure of air is maintained to ensure uniform thickness of the film bubble.

The film bubble is cooled below the softening point of the polymer by blowing air on it from a cooling ring placed round the die. When the polymer, such as polyethylene, cools below the softening point, the crystalline material is cloudy compared with the clear amorphous melt. The transition line which coincides with this transformation is therefore called the frost line.

The ratio of bubble diameter to die diameter is called the blowup ratio. It may range as high as 4 or 5, but 2.5 is a more typical figure. Molecular orientation occurs in the film in the hoop direction during blowup, and orientation in the machine direction, that is, in the direction of the extrudate flow from the die, can be induced by tension from the pinch rolls. The film bubble after solidification (at frost line) moves upward through guiding devices into a set of pinch rolls which flatten it. It can then be slit, gusseted, and surface-treated in line. (Vertical extrusion, shown in Figure 1.23, is most common, although horizontal techniques have been successfully used.)

Blown-film extrusion is an extremely complex subject, and a number of problems are associated with the production of good-quality film. Among the likely defects are variation in film thickness, surface

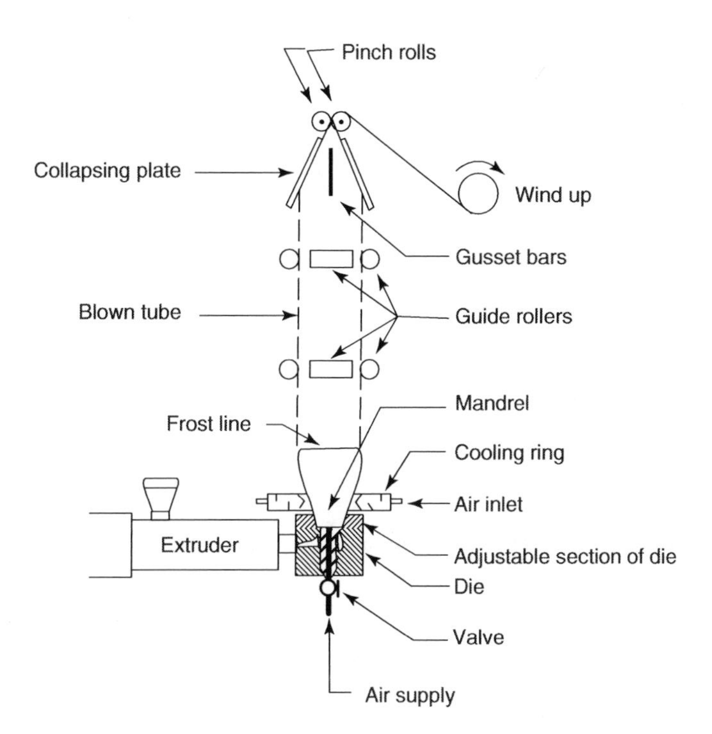

FIGURE 1.23 Typical blow-film extrusion setup.

imperfections (such as "fish eyes," "orange peel," haze), wrinkling, and low tensile strength. The factors affecting them are also numerous. "Fish eyes" occur due to imperfect mixing in the extruder or due to contamination of the molten polymer. Both factors are controlled by the screen pack.

The blown-film technique has several advantages: the relative ease of changing film width and caliber by controlling the volume of air in the bubble and the speed of the screw; the elimination of the end effects (e.g., edge bead trim and nonuniform temperature that result from flat film extrusion); and the capability of biaxial orientation (i.e., orientation both in the hoop direction and in the machine direction), which results in nearly equal physical properties in both directions, thereby giving a film of maximum toughness.

After extrusion, blown-film is often slit and wound up as flat film, which is often much wider than anything produced by slot-die extrusion. Thus, blown-films of diameters 7 ft. or more have been produced, giving flat film of widths up to 24 ft. One example is reported [16] of a 10-in. extruder with 5-ft diameter and a blowup ratio of 2.5, producing 1,100 lb/h of polyethylene film, which when collapsed and slit is 40 ft wide. Films in thicknesses of 0.004–0.008 in. are readily produced by the blown-film process. Polyethylene films of such large widths and small thicknesses find extensive uses in agriculture, horticulture, and building.

1.7.5 Flat Film or Sheet Extrusion

In the flat-film process the polymer melt is extruded through a slot die (T-shaped or "coat hanger" die), which may be as wide as 10 ft. The die has relatively thick wall sections on the final lands (as compared to the extrusion coating die) to minimize deflection of the lips from internal melt pressure. The die opening (for polyethylene) may be 0.015–0.030 in.—even for films that are less than 0.003 in. thick. The reason is that the speed of various driven rolls used for taking up the film is high enough to draw down the film with a concurrent thinning. (By definition, the term film is used for material less than 0.010 in. thick, and *sheet* for that which in thicker.)

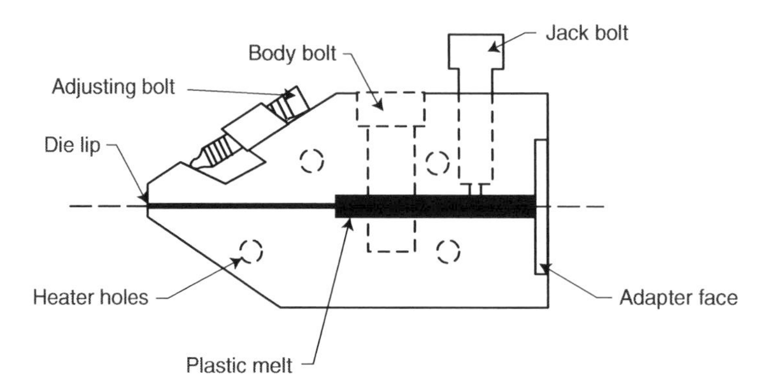

FIGURE 1.24 Sheet extrusion die.

Figure 1.24 illustrates the basic components of a sheet extrusion die. The die is built in two halves the part for easier construction and maintenance. The use of a jack bolt facilitates separation of the die halves when the die is full of plastic. The die lip can be adjusted (across the entire lip length) to enable the processor to keep the thickness of the extruded sheet within specification.

Figure 1.25 illustrates a T-type die and a coat-hanger-type die, which are used for both film and sheet extrusion. The die must produce a smooth and uniform laminar flow of the plastic melt which has already been mixed thoroughly in the extruder. The internal shape of the die and the smoothness of the die surface are critical to this flow transition. The deckle rods illustrated in Figure 1.25 are used by the processor to adjust the width of the extruded sheet or film.

Following extrusion, the film may be chilled below T_m or T_g by passing in through a water bath or over two or more chrome-plated chill rollers which have been cored for water cooling. A schematic drawing of a chill-roll (also called cast-film) operation is shown in Figure 1.26. The polymer melt extruded as a web from the die is made dimensionally stable by contacting several chill rolls before being pulled by the powered carrier rolls and wound up. The chrome-plated surface of the first roll is highly polished so that the product obtained is of extremely high gloss and clarity.

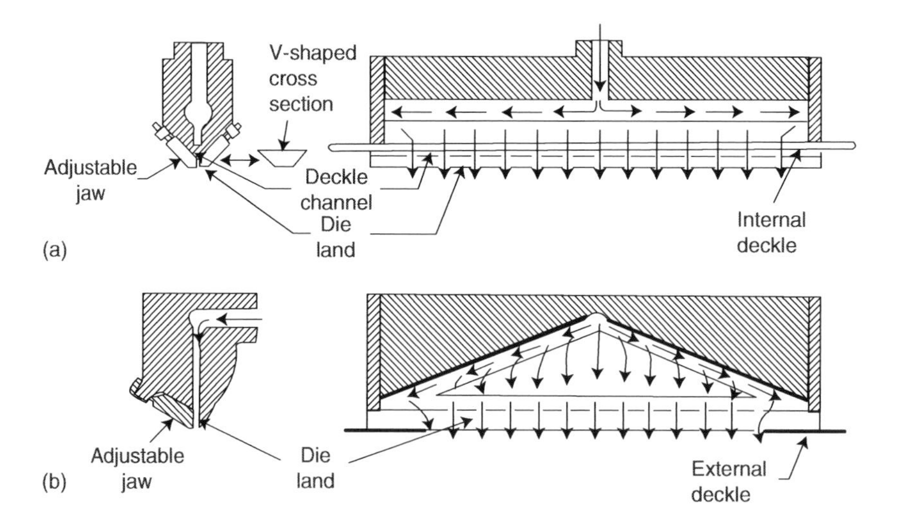

FIGURE 1.25 Schematic cross-sections (a) T-type and (b) coat-hanger-type extrusion dies. (After *Petrothene: A Processing Guide*, 3rd Ed., 1965. U.S. Industrial Chemicals Co., New York.)

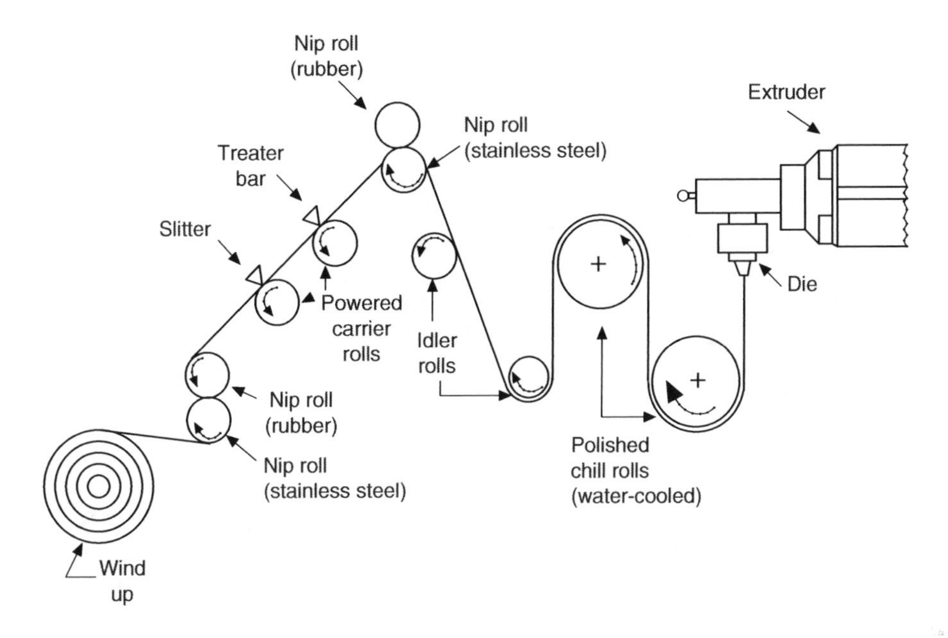

FIGURE 1.26 Sketch of chill-roll film extrusion. (After *Petrothene: A Processing Guide*, 3rd Ed., 1965. U.S. Industrial Chemicals Co., New York.)

In flat-film extrusion (particularly at high takeoff rates), there is a relatively high orientation of the film in the machine direction (i.e., the direction of the extrudate flow) and a very low one in the traverse direction.

Biaxially oriented film can be produced by a flat-film extrusion by using a tenter (Figure 1.27). Polystyrene, for example, is first extruded through a slit die at about 190°C and cooled to about 120°C by passing between rolls. Inside a temperature-controlled box and moving sheet, rewarmed to 130°C, is grasped on either side by tenterhooks which exert a drawing tension (longitudinal stretching) as well as widening tension (lateral stretching). Stretch ratios of 3:1–4:1 in both

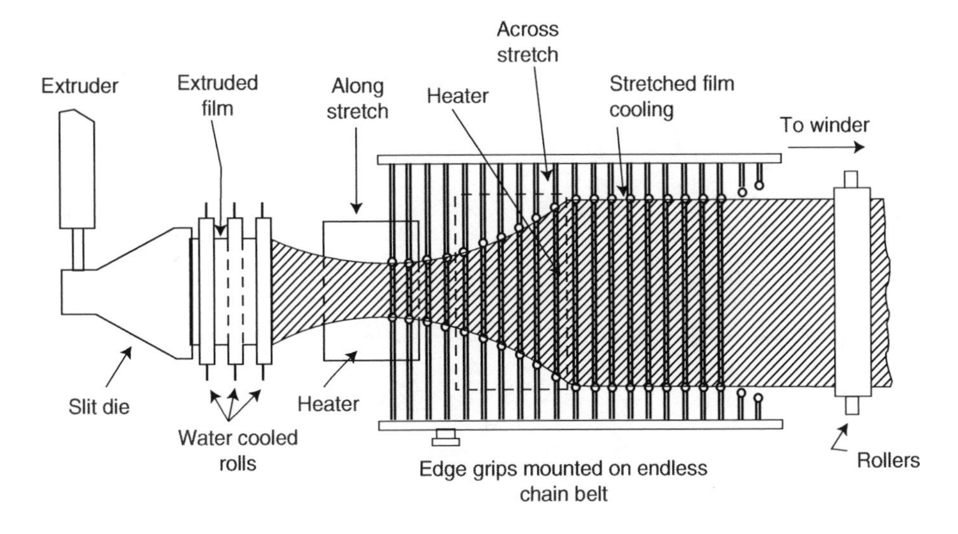

FIGURE 1.27 Plax process for manufacture of biaxially stretched polystyrene film. (After Brydson, J.A. 1982. *Plastics Materials*, Butterworth Scientific, London, U.K.)

directions are commonly employed for biaxially oriented polystyrene film. Biaxial stretching leads to polymers of improved tensile strength. Commercially available oriented polystyrene film has a tensile strength of 10,000–12,000 psi (703–843 kg/cm^2), compared to 6,000–8,000 psi (422–562 kg/cm^2) for unstretched material.

Biaxial orientation effects are important in the manufacture of films and sheet. Biaxially stretched polypropylene, poly(ethyleneterephthalate) (e.g., Melinex) and poly(vinylidene chloride) (Saran) produced by flat-film extrusion and tentering are strong films of high clarity. In biaxial orientation, molecules are randomly oriented in two dimensions just as fibers would be in a random mat; the orientation-induced crystallization produces structures which do not interfere with the light waves. With polyethylene, biaxial orientation often can be achieved in blown-film extrusion.

1.7.6 Pipe or Tube Extrusion

The die used for the extrusion of pipe or tubing consists of a die body with a tapered mandrel and an outer die ring which control the dimensions of the inner and outer diameters, respectively. Since this process involves thicker walls than are involved in blown-film extrusion, it is advantageous to cool the extrudate by circulating water through the mandrel (Figure 1.28) as well as by running the extrudate through a water bath.

The extrusion of rubber tubing, however, differs from thermoplastic tubing. For thermoplastic tubing, dimensional stability results from cooling below T_g or T_m, but rubber tubing gains dimensional stability due to a cross-linking reaction at a temperature above that in the extruder. The high melt viscosity of the rubber being extruded ensures a constant shape during the cross-linking.

A complication encountered in the extrusion of continuous shapes is die swell. Die swell is the swelling of the polymer when the elastic energy stored in capillary flow is relaxed on leading the die. The extrusion of flat sheet or pipe is not sensitive to die swell, since the shape remains symmetrical even through the dimensions of the extrudate differ from those of the die. Unsymmetrical cross sections may, however, be distorted.

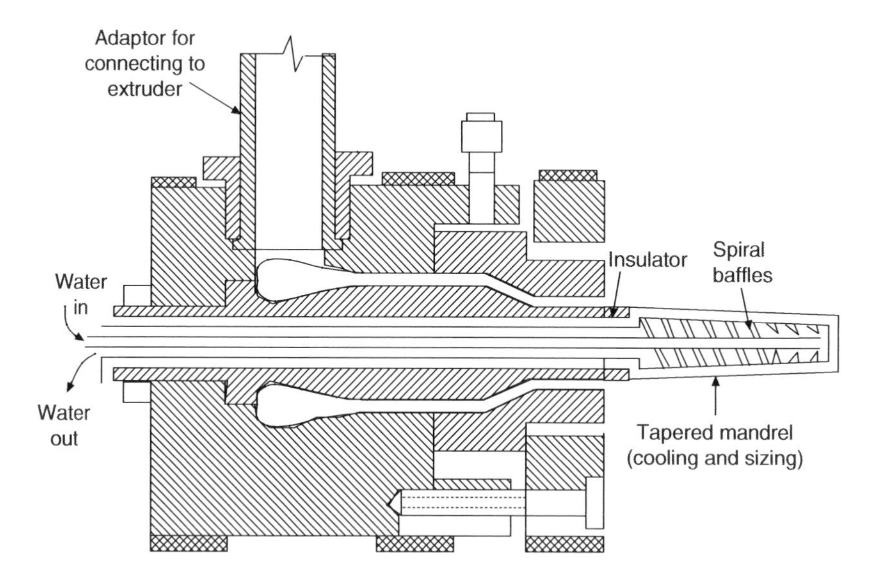

FIGURE 1.28 An extrusion die fitted with a tapered cooling and sizing mandrel for use in producing either pipe or tubing. (After Van Ness, R.T., De Hoff, G.R., and Bonner, R.M. 1968. *Mod. Plastics*, 45, 14A, Encycl. Issue, 672.)

1.7.7 Wire and Cable Coverings

The covering or coating of wire and cable in continuous lengths with insulating plastics is an important application of extrusion, and large quantities of resin are used annually for this purpose. This applications represented one of the first uses of extruders for rubber about 100 years ago. The wire and cable coating process resembles the process used for pipe extrusion (Figure 1.28) with the difference that the conductor (which may be a single metal strand, a multiple strand, or even a bundle of previously individually insulated wires) to be covered is drawn through the mandrel on a continuous basis (Figure 1.29). For thermoplastics such as polyethylene, nylon, and plasticized PVC, the coating is hardened by cooling below T_m or T_g by passing through a water trough. Rubber coatings, on the other hand, are to be cross-linked by heating subsequent to extrusion.

1.7.8 Extrusion Coating

Many substrates, including paper, paperboard, cellulose film, fireboard, metal foils, or transparent films are coated with resins by direct extrusion. The resins most commonly used are the polyolefins, such as polyethylene, polypropylene, ionomer, and ethylene–vinyl acetate copolymers. Nylon, PVC, and polyester are used for a lesser extent. Often combinations of these resins and substrates are used to provide a multiplayer structure. [A related technique, called extrusion laminating, involves two or more substrates, such as paper and aluminum foil, combined by using a plastic film (e.g., polyethylene), as the adhesive and as a moisture barrier.] Coatings are applied in thicknesses of about 0.2–15 mils, the common average being 0.5–2 mils, and the substrates range in thickness from 0.5 to more than 24 mils.

The equipment used for extrusion coating is similar to that used for the extrusion of flat film. Figure 1.30 shows a typical extrusion coating setup. The thin molten film from the extruder is pulled down into the nip between a chill roll and a pressure roll situated directly below the die. The pressure between these two rolls forces the film on to the substrate while the substrate, moving at a speed faster than the extruded film, draws the film to the required thickness. The molten film is cooled by the water-cooled, chromium-plated chill roll. The pressure roll is also metallic but is covered with a rubber sleeve, usually neoprene or silicone rubber. After trimming, the coated material is wound up on conventional windup equipment.

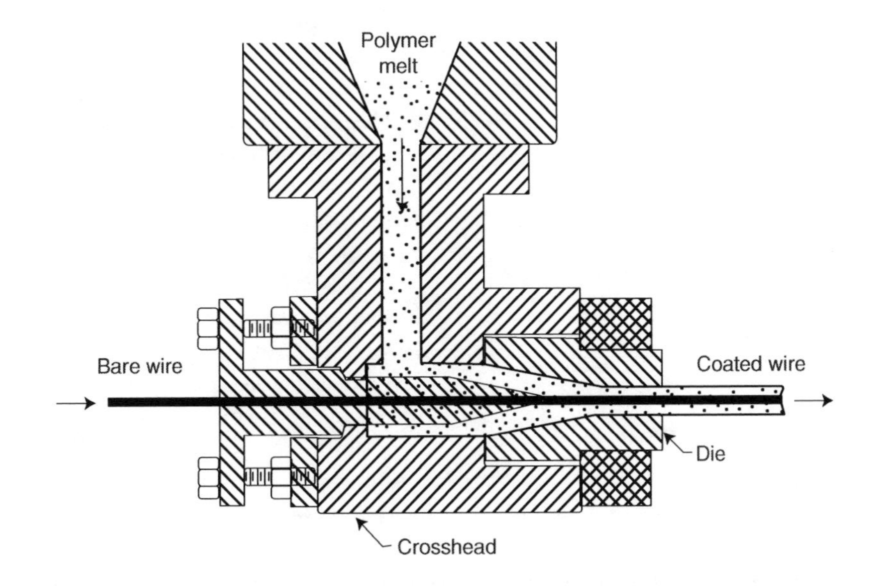

FIGURE 1.29 Crosshead used for wire coating.

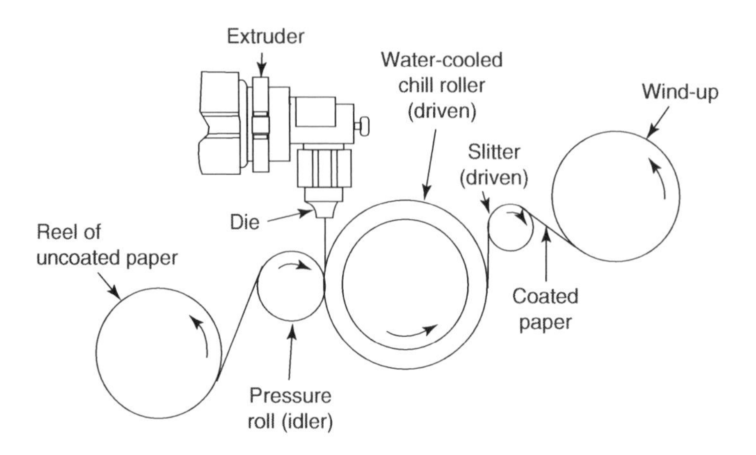

FIGURE 1.30 Sketch of paper coating for extrusion process.

1.7.9 Profile Extrusion

Profile extrusion is similar to pipe extrusion (Figure 1.28) except that the sizing mandrel is obviously not necessary. A die plate, in which an orifice of appropriate geometry has been cut, is placed on the face of the normal die assembly. The molten polymer is subjected to surface drag as it passes through the die, resulting is reduced flow through the thinner sections of the orifice. This effect is countered by altering the shape of the orifice, but often this results in a wide difference in the orifice shape from the desired extrusion profile. Some examples are shown in Figure 1.31.

1.8 Blow Molding

Basically, blow molding is intended for use in manufacturing hollow plastic products, such as bottles and other containers [17]. However, the process is also used for the production of toys, automobile parts, accessories, and many engineering components. The principles used in blow molding are essentially similar to those used in the production of glass bottles. Although there are considerable differences in the process available for blow molding, the basic steps are the same: (1) melt the plastic; (2) form the molten plastic into a *parison* (a tubelike shape of molten plastic); (3) seal the ends of the parison except for one area through which the blowing air can enter; (4) inflate the parison to assume the shape of the mold in which it is placed; (5) cool the blow-molded part; (6) eject the blow-molded part; (7) trim flash if necessary.

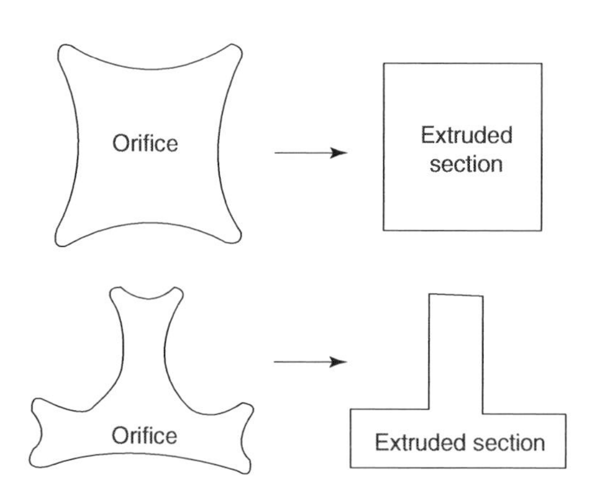

FIGURE 1.31 Relationships between extruder die orifice and extruded section.

Two basic processes of blow molding are extrusion blow molding and injection blow molding. These processes differ in the way in which the parison is made. The extrusion process utilizes an unsupported parison, whereas the injection process utilizes a parison supported on a metal core. The extrusion blow-molding process by far accounts for the largest percentage of blow-molded objects produced today. The injection process is, however, gaining acceptance.

Although any thermoplastic can be blow-molded, polyethylene products made by this technique are predominant. Polyethylene squeeze bottles form a large percentage of all blow-molded products.

1.8.1 Extrusion Blow Molding

Extrusion blow molding consists basically of the extrusion of a predetermined length of parison (hollow tube of molten plastic) into a split die, which is then closed, sealing both ends of the parison. Compressed air is introduced (through a blowing tube) into the parison, which blows up to fit the internal contours of the mold. As the polymer surface meets the cold metal wall of the mold, it is cooled rapidly below T_g or T_m. When the product is dimensionally stable, the mold is opened, the product is ejected, a new parison is introduced, and the cycle is repeated. The process affords high production rates.

In continuous extrusion blow molding, a molten parison is produced continuously from a screw extruder. The molds are mounted and moved. In one instance the mold sets are carried on a rotating horizontal table (Figure 1.32a), in another on the periphery of a rotating vertical wheel (Figure 1.32b). Such rotary machines are best suited for long runs and large-volume applications.

In the ram extrusion method the parison is formed in a cyclic manner by forcing a charge out from an accumulated molten mass, as in the preplasticizer injection-molding machine. The transport arm cuts and holds the parison and lowers it into the waiting mold, where shaping under air pressure takes place (Figure 1.33).

A variation of the blow-extrusion process which is particularly suitable for heat-sensitive resins such as PVC is the cold preform molding. The parison is produced by normal extrusion and cooled and stored until needed. The required length of tubing is then reheated and blown to shape in a cold mold, as in conventional blow molding. Since, unlike in the conventional process, the extruder is not coupled directly to the blow-molding machine, there is less chance of a stoppage occurring, with consequent risk of holdup and degradation of the resin remaining in the extruder barrel. There is also less chance of the occurrence of "dead" pockets and consequent degradation of resin in the straight-through die used in this process than in the usual crosshead used with a conventional machine.

1.8.2 Injection Blow Molding

In this process the parison is injection molded rather than extruded. In one system, for example, the parison is formed as a thick-walled tube around a blowing stick in a conventional injection-molding machine. The parison is then transferred to a second, or blowing, mold in which the parison is inflated to the shape of the mold by passing compressed air down the blowing stick. The sequence is shown in

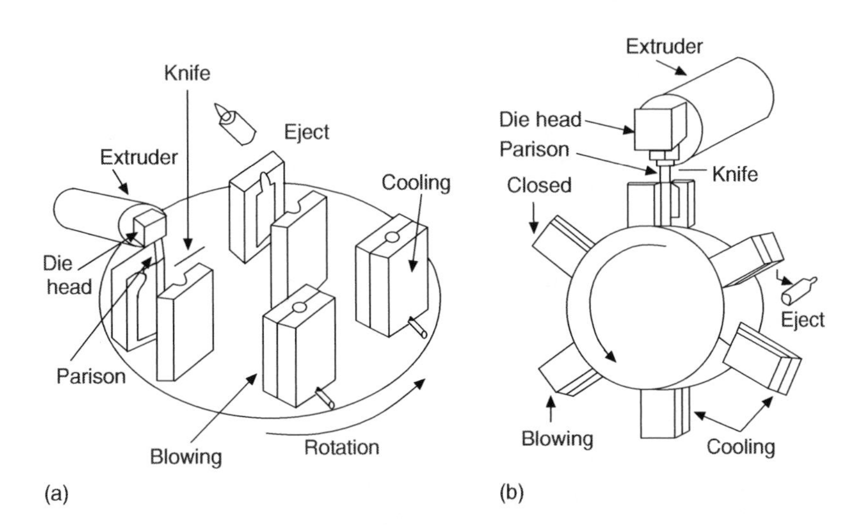

FIGURE 1.32 Continuous extrusion blow molding. (a) Rotating horizontal table carrying mold sets. (b) Continuous vertical rotation of a wheel carrying mold sets on the periphery. (After Morgan, B.T., Peters, D.L., and Wilson, N.R. 1967. *Mod. Plastics*, 45, 1A, Encycl. Issue, 797.)

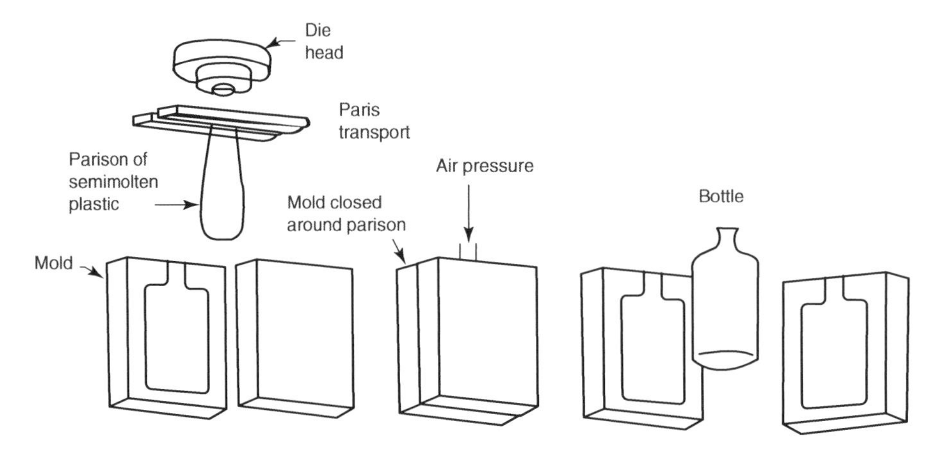

FIGURE 1.33 Continuous extrusion blow molding with parison transfer. The transport arm cuts the extruded parison from the die head and lowers it into the waiting mold. (After Morgan, B.T., Peters, D.L., and Wilson, N.R. 1967. *Mod. Plastics*, 45, 1A, Encycl. Issue, 797.)

Figure 1.34. Injection blow molding is relatively slow and is more restricted in choice of molding materials as compared to extrusion blow molding. The injection process, however, affords good control of neck and wall thicknesses of the molded object. With this process it is also easier to produce unsymmetrical molding.

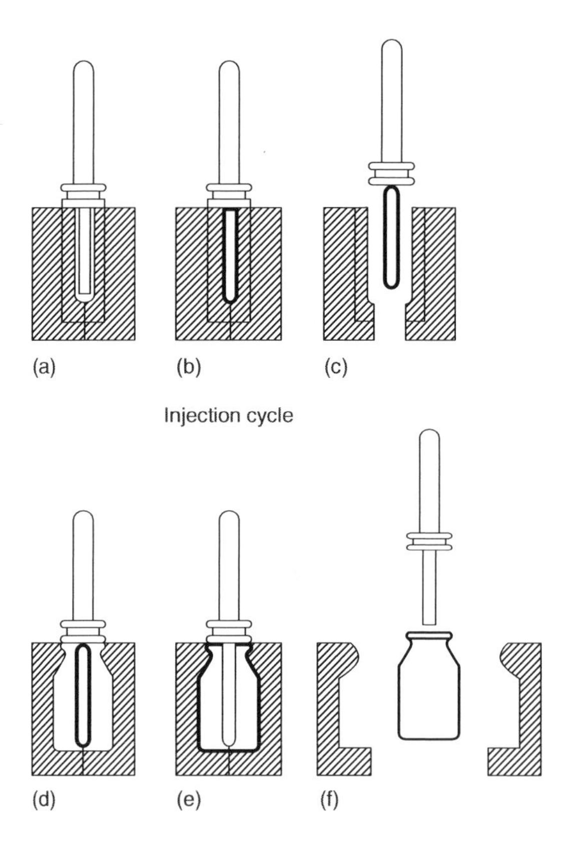

FIGURE 1.34 Sketch of injection-blow-molding process.

1.8.3 Blow Molds

Generally, the blow mold is a cavity representing the outside of a blow-molded part. The basic structure of a blow mold consists of a cast or machined block with a cavity, cooling system, venting system, pinchoffs, flash pockets, and mounting plate. The selection of material for the construction of a blow mold is based on the consideration of such factors as thermal conductivity, durability, cost of the material, the resin being processed, and the desired quality of the finished parts. Commonly used mold materials are beryllium, copper, aluminum, ampcoloy, A-2 steel, and 17-4 and 420 stainless steels.

Beryllium–copper (BeCu) alloys 165 and 25 are normally used for blow molds. These materials display medium to good thermal conductivity with good durability. Stainless steels such as 17-4 and 420 are also frequently employed in blow molds where durability and resistance to hydrochloric acid are required. Heat-treated A-2 steel is often used as an insert in pinchoffs where thermal conductivity is not a concern and high quality parts are required.

For blow molding HDPE parts, aluminum is commonly employed for the base material, with BeCu or stainless steel inserts in the pinchoff areas. For PVC parts BeCu, ampcoloy, or 17-4 stainless steel are used as the base material, with A-2 or stainless inserts in the pinchoff area. For PET parts, the base mold is typically made of aluminum or BeCu, with A-2 or stainless steel pinchoffs.

The production speed of blow-molded parts is generally limited by one of two factors: extruder capacity or cooling time in the mold. Cooling of mold is accomplished by a water circuit into the mold. Flood cooling and cast-in tubes are most common in cast molds; drilled holes and milled slots are the norm in machined blow molds.

In multiple-cavity molds, series and parallel cooling circuits are used. Series cooling enters and cools one cavity, then moves to the next until all the cavities are cooled. The temperature of the water increases as it moves through the mold, and this results in non-uniform cooling. Parallel cooling, on the other hand, enters and exists all cavities simultaneously, thereby cooling all cavities at a uniform rate. Parallel cooling is thus the preferred method but it is not always possible due to limitations.

1.9 Calendering

Calendering is the leading method for producing vinyl film, sheet, and coatings [18]. In this process continuous sheet is made by passing a heat-softened material between two or more rolls. Calendering was originally developed for processing rubber, but is now widely used for producing thermoplastic films, sheets, and coatings. A major portion of thermoplastics calendered is accounted for by flexible (plasticized) PVC. Most plasticized PVC film and sheet, ranging from the 3-mil film for baby pants to the 0.10 in "vinyl" tile for floor coverings, is calendered.

The calendaring process consists of feeding a softened mass into the nip between two rolls where it is squeezed into a sheet, which then passes round the remaining rolls. The processed material thus emerges as a continuous sheet, the thickness of which is governed by the gap between the last pair of rolls. The surface quality of the sheet develops on the last roll and may be glossy, matt, or embossed. After leaving the calender, the sheet is passed over a number of cooling rolls and then through a beta-ray thickness gage before being wound up.

The plastics mass fed to the calender may be simply a heat-softened material, as in the case of, say, polyethylene, or a rough sheet, as in the case of PVC. The polymer PVC is blended with stabilizers, plasticizers, etc., in ribbon blenders, gelated at 120°C–160°C for about 5–10 min in a Banbury mixer, and the gelated lumps are made into a rough sheet on a two-roll mill before being fed to the calender.

Calenders may consist of two, three, four, or five hollow rolls arranged for steam heating or water cooling and are characterized by the number of rolls and their arrangement. Some arrangements are shown in Figure 1.35. Thick sections of rubber can be made by applying one layer of polymer upon a previous layer (double plying) (Figure 1.35b). Calenders can be used for applying rubber or plastics to fabrics (Figure 1.35c). Fabric or paper is fed through the last two rolls of the calender so that the resin film is pressed into the surface of the web. For profiling, the plastic material is fed to the nip of the calender,

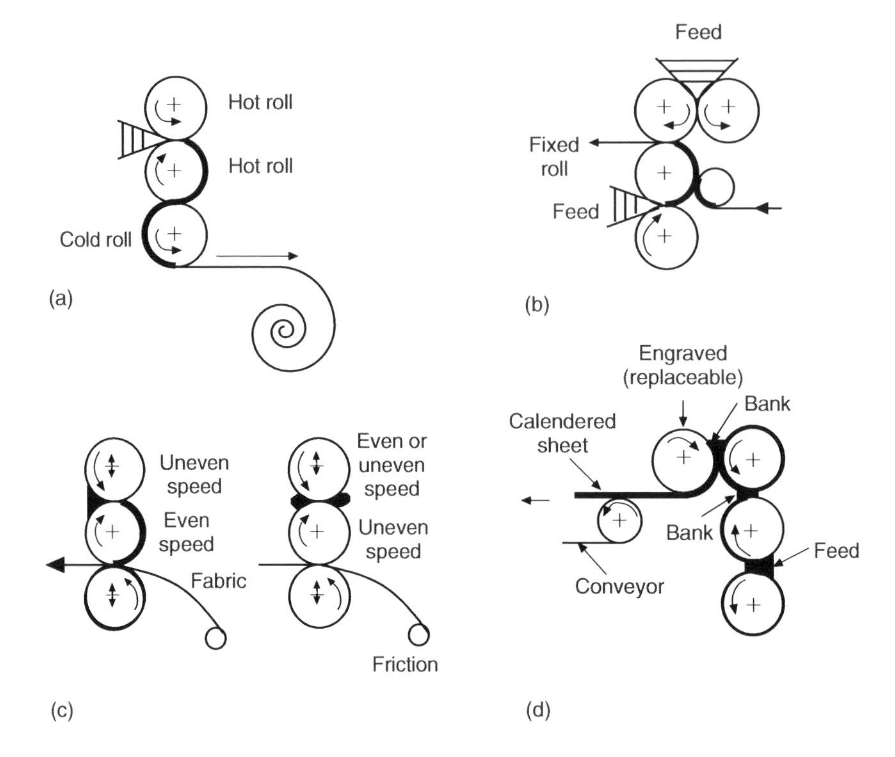

FIGURE 1.35 Typical arrangements of calender rolls: (a) single-ply sheeting; (b) double-ply sheeting; (c) applying rubber to fabrics; (d) profiling with four-roll engraving cylinder.

where the material assumes the form of a sheet, which is then progressively pulled through two subsequent banks to resurface each of the two sides (Figure 1.35d). For thermoplastics the cooling of the sheet can be accomplished on the rolls with good control over dimensions. For rubber, cross-linking can be carried out with good control over dimensions, with the support of the rolls. Despite the simple appearance of the calender compared to the extruder, the close tolerances involved and other mechanical problems make for the high cost of a calendaring unit.

1.10 Spinning of Fibers

The term spinning, as used with natural fibers, refers to the twisting of short fibers into continuous lengths [19–21]. In the modern synthetic fiber industry, however, the term is used for any process of producing continuous lengths by any means. (A few other terms used in the fiber industry should also be defined. A fiber may be defined as a unit of matter having a length at least 100 times its width or diameter. An individual strand of continuous length is called a filament. Twisting together filaments into a strand gives continuous filament yarn. If the filaments are assembled in a loose bundle, we have tow or roving. These can be chopped into small lengths (an inch to several inches long), referred to as staple. Spun yarn is made by twisting lengths of staple into a single continuous strand, and *cord* is formed by twisting together two or more yarns.)

The dimensions of a filament or yarn are expressed in terms of a unit called the "tex" which is a measure of the fineness or linear density. One tex is 1 gram per 1,000 meters or 10^{-6} kg/m. The tex has replaced "denier" as a measure of the density of the fiber. One denier is 1 gram per 9,000 meters, so 1 denier = 0.1111 tex.

The primary fabrication process in the production of synthetic fibers is the spinning—i.e., the formation—of filaments. In every case the polymer is either melted or dissolved in a solvent and is put in filament form by forcing through a die, called spinneret, having a multiplicity of holes. Spinnerets for rayon spinning, for example, have as many as 10,000 holes in a 15-cm-diameter platinum disc, and those for textile yarns may have 10–120 holes; industrial yarns such as tire core might be spun from spinnerets with up to 720 holes.

Three major categories of spinning processes are melt, dry, and wet spinning [19]. The features of the three processes are shown in Figure 1.36, and the typical cross sections of the fibers produced by them are shown in Figure 1.37.

1.10.1 Melt Spinning

In melt spinning, which is the same as melt extrusion, the polymer is heated and the viscous melt is pumped through a spinneret. An inert atmosphere is provided in the melting chamber before the pump. Special pumps are used to operate in the temperature range necessary to produce a manageable melt (230–315°C). For nylon, for example, a gear pump is used to feed the melt to the spinneret (Figure 1.36a). For a polymer with high melt viscosity such as polypropylene, a screw extruder is used to feed a heated spinneret. Dimensional stability of the fiber is obtained by cooling under tension.

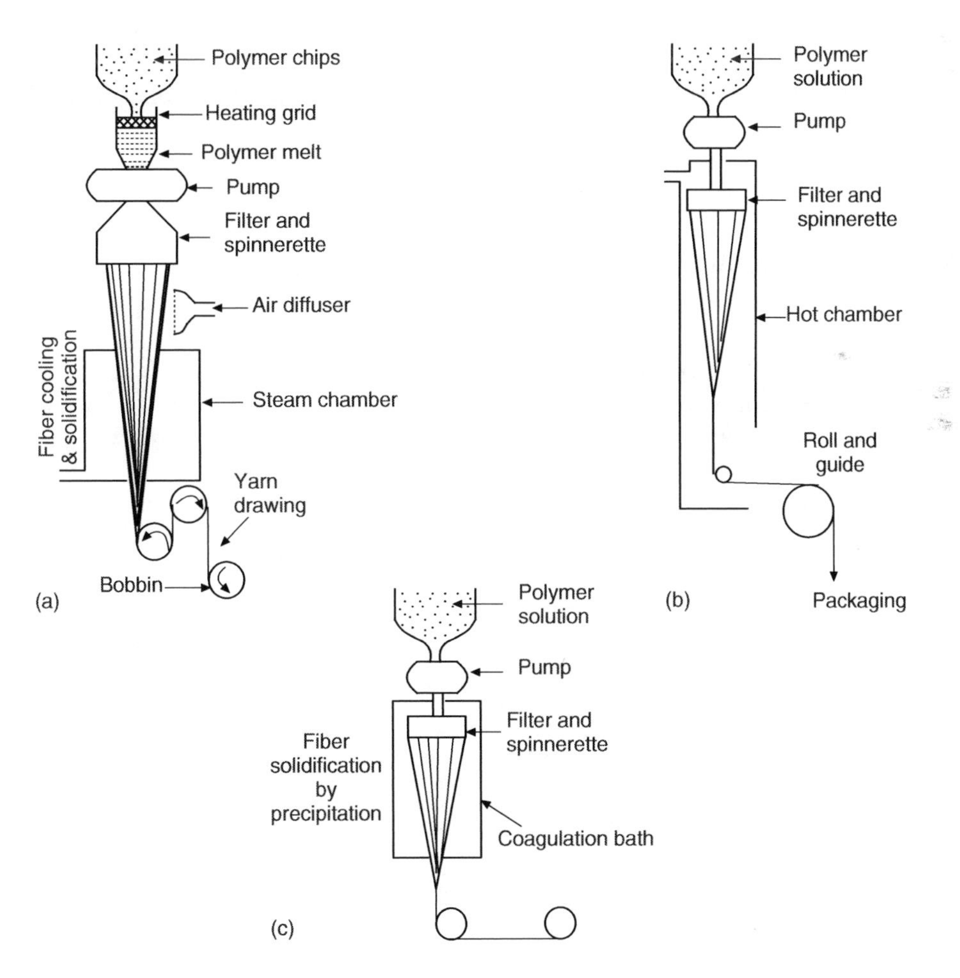

FIGURE 1.36 Schematic of the three principal types of fiber spinning: (a) melt spinning; (b) dry spinning; (c) wet spinning. (After Carraher, C. E., Jr. 2002. *Polymer News*, 27, 3, 91.)

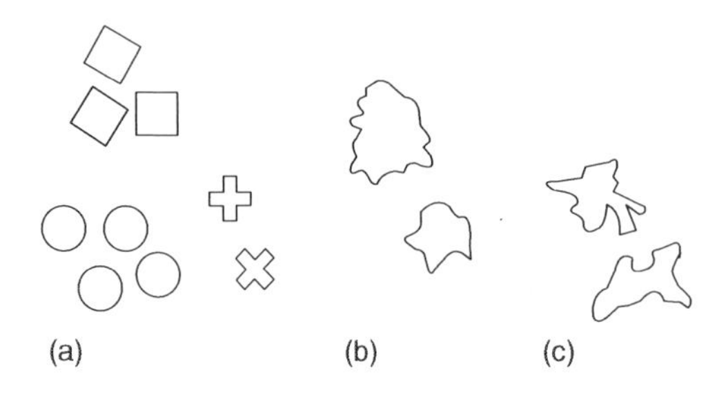

FIGURE 1.37 Typical cross-section of fibers produced by different spinning processes: (a) melt-spun nylon from various shaped orifices; (b) dry-spun cellulose acetate from round orifice; (c) wet-spun viscose rayon from round orifice.

Typical melt spinning temperatures are given in Table 1.1.

1.10.2 Dry Spinning

In dry spinning, a polymer is dissolved in a solvent and the polymer solution (concentration on the order of 20–40%) is filtered and then forced through a spinneret into a chamber through which heated air is passed to achieve dimensional stability of the fiber by evaporation of the solvent (Figure 1.36b). For economical reasons, the gas is usually air, but inert gases such as nitrogen and superheated steam are sometimes used. The skin which forms first on the fiber by evaporation from the surface gradually collapses and wrinkles as more solvent diffuses out and the diameter decreases. The cross section of a dry-spun fiber thus has an irregularly lobed appearance (Figure 1.37). Recovery of the solvent used for dissolving the polymer is important to the economics of the process. Cellulose acetate dissolved in acetone and polyacrylonitrile (PAN) dissolved in dimethylformamide are two typical examples.

The hot solution (dope) of PAN in DMF is extruded directly into a hot stream of nitrogen at 300°C. The residual DMF is recovered in subsequent water washing steps.

Dry spun fibers have lower void concentrations than wet spun fibers. This is reflected in greater densities and lower dyeability for the dry spun fibers.

1.10.3 Wet Spinning

Wet spinning also involves pumping a solution of the polymer to the spinneret. However, unlike dry spinning, dimensional stability is achieved by precipitating the polymer in a nonsolvent (Figure 1.36c). For example, PAN in dimethylformamide can be precipitated by passing a jet of the solution through a bath of water, which is miscible with the solvent but coagulates the polymer. For wet-spinning cellulose triacetate a mixture methylene chloride and alcohol can be used to dissolve the polymer, and a toluene bath can be used for precipitation of the polymer. In some cases the precipitation can also involve a chemical reaction. An important example is viscose rayon, which is made by regenerating cellulose from a solution of cellulose xanthate in dilute alkali.

$$\mathrm{R{-}OH} \xrightarrow{\mathrm{CS_2,\ NaOH}} \underset{\text{cellulose xanthate}}{\mathrm{R{-}O{-}\overset{\overset{\displaystyle S}{\|}}{C}{-}S^-\ Na^+}} + \mathrm{H_2O}$$

$$\Big\downarrow\ \mathrm{H_2O};\ \mathrm{H_2SO_4},\ \mathrm{NaHSO_4}$$

$$\underset{\text{cellulose}}{\mathrm{R{-}OH}} + \mathrm{CS_2} + \mathrm{Na^+\ (HSO_4^-)}$$

If a slot die rather than a spinneret is used, the foregoing process would yield cellulose film (cellophane) instead of fiber.

TABLE 1.1 Typical Spinning Temperatures for Selected Polymers

Polymer	Melting Point (°C)	Typical Spinning Temperature (°C)
Nylon-6	220	280
Nylon-6,6	260	290
Poly(ethylene terephthalate)	260	290
Polyethylene	~ 130	220–230
Polypropylene	170	250–300
Poly(vinylidene chloride) copolymers	120–140	180

Source: Carraher, C. E. Jr. 2002. *Polymer News*, 27(3), 91.

Wet spinning is the most complex of the three spinning processes, typically including washing, stretching, drying, crimping, finish application, and controlled relaxation to form tow material [22]. A simplified sketch of the Asahi wet spinning process for making PAN (acrylic) fiber tows is shown in Figure 1.38. The polymer solution (dope) is made in concentrated HNO_3 (67%) at low temperatures using pulverizer and mixer, filtered, and deaerated. The dope, containing 14–15% polymer and maintained at -7°C, is extruded at a pumping pressure of 10–15 atm through the spinnerets immersed in the coagulation bath. Each spinneret has about 46,000–73,000 holes (different sizes for different grades). The acid concentration in the bath is 37% and the temperature is -5°C. The filaments from 5 spinnerets are collected into a tow. The spinning speed in this step is about 7 meters per minute. The dilute nitric acid from the bath goes to the concentration section where 67% HNO_3 is obtained for re-use in dope preparation.

In the next pre-finishing step, the fibers (tows) are repeatedly washed with water by spraying and immersion to remove the acid. The fiber is stretched in three stages, a 1:10 stretch being obtained in the last hot water bath at 100°C. In the finishing section, the tows pass through a water bath containing 1% oil to impart antistaticity and reduce friction. The tows are then dried in a hot air (135°C) dryer over rollers. Subsequent treatments include a second finishing oil spray, crimping, second hot air drying, plaiting, thermosetting, and cutting into staple fibers.

Fibers made from wet spinning generally have high void contents in comparison to all of the other processes giving them increased dyeability and the surface is rougher with longitudinal serrations.

Hollow fibers for gas and liquid separation are prepared by passing air through the material just prior to entrance into the non-solvent bath.

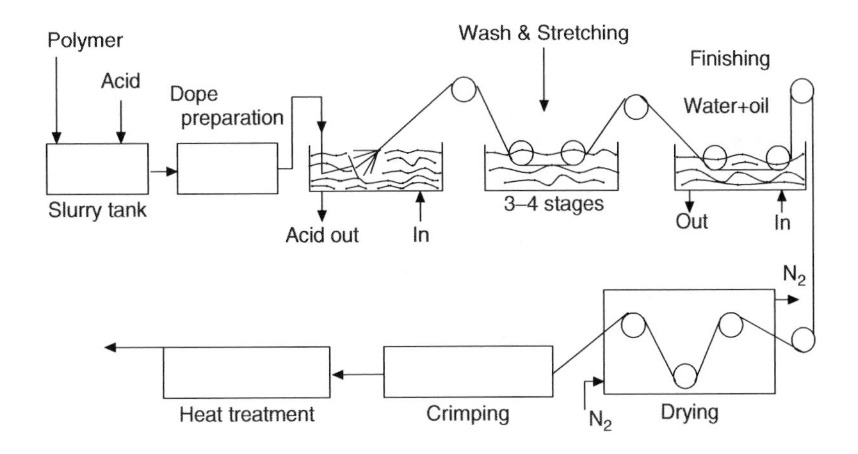

FIGURE 1.38 A simplified sketch of the Asahi wet spinning process for polyacrylonitrile fiber.

1.10.4 Cold Drawing of Fibers

Almost all synthetic fibers are subjected to a drawing (stretching) operation to orient the crystalline structure in the direction of the fiber axis. Drawing orients crystallites in the direction of the stretch so that the modulus in that direction is increased and elongation at break is decreased. Usually the drawing is carried out at a temperature between T_g and T_m of the fiber. Thus, polyethylene ($T_g = -115°C$) can be drawn at room temperature, whereas nylon−6,6 ($T_g = 53°C$) should be heated or humidified to be drawn. T_g is depressed by the presence of moisture, which acts as a plasticizer. The drawing is accomplished by winding the yarn around a wheel or drum driven at a faster surface velocity than a preceding one.

1.11 Thermoforming

When heated, thermoplastic sheet becomes as soft as a sheet of rubber, and it can then be stretched to any given shape [23]. This principle is utilized in thermoforming processes which may be divided into three main types: (a) vacuum forming, (2) pressure forming (blow forming), and (3) mechanical forming (e. g., matched metal forming), depending on the means used to stretch the heat softened sheet.

Since fully cured thermoset sheets cannot be resoftened, forming is not applicable to them. Common materials subjected to thermoforming are thermoplastics such as polystyrene, cellulose acetate, cellulose acetate butyrate, PVC, ABS, poly(methyl methacrylate), low- and high-density polyethylene, and polypropylene. The bulk of the forming is done with extruded sheets, although cast, calendered, or laminated sheets can also be formed.

In general, thermoforming techniques are best suited for producing moldings of large area and very thin-walled moldings, or where only short runs are required. Thermoformed articles include refrigerator and freezer door liners complete with formed-in compartments for eggs, butter, and bottles of various types, television masks, dishwasher housings, washing machine covers, various automobile parts (instrument panels, arm rests, ceilings, and door panels), large patterned diffusers in the lighting industry, displays in advertising, various parts in aircraft industry (windshields, interior panels, arm rests, serving trays, etc.), various housing (typewriters, dictaphones, and duplicating machines), toys, transparent packages, and much more.

1.11.1 Vacuum Forming

In vacuum forming, the thermoplastic sheet can be clamped or simply held against the RIM of a mold and then heated until it becomes soft. The soft sheet is then sealed at the RIM, and the air from the mold cavity is removed by a suction pump so that the sheet is forced to take the contours of the mold by the atmospheric pressure above the sheet (Figure 1.39a). The vacuum in the mold cavity is maintained until the part cools and becomes rigid.

Straight cavity forming is not well adapted to forming a cup or box shape because as the sheet, drawn by vacuum, continues to fill out the mold and solidify, most of the stock is used up before it reaches the periphery of the base, with the result that this part becomes relatively thin and weak. This difficulty is alleviated and uniformity of distribution in such shapes is promoted if the *plug assist* is used (Figure 1.39b). The plug assist is any type of mechanical helper which carries extra stock toward an area where the part would otherwise be too thin.

Plug-assist techniques are adaptable both to vacuum-forming and pressure forming techniques. The system shown in Figure 1.39b is thus known as plug assist vacuum forming.

1.11.2 Pressure Forming

Pressure forming is the reverse of vacuum forming. The plastic sheet is clamped, heated until it becomes soft, and sealed between a pressure head and the RIM of a mold. By applying air pressure (Figure 1.40),

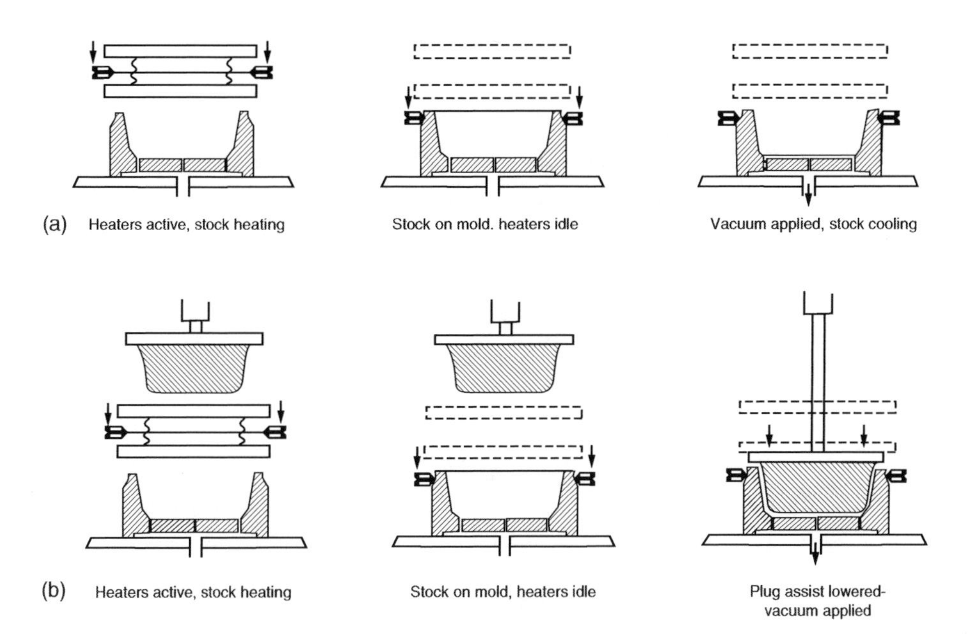

FIGURE 1.39 (a) Vacuum forming. (b) Plug-assist forming using vacuum.

one forces the sheet to take the contours of the mold. Exhaust holes in the mold allow the trapped air to escape. After the part cools and becomes rigid, the pressure is released and the part is removed. As compared to vacuum forming, pressure forming affords a faster production cycle, greater part definition, and greater dimensional control.

A variation of vacuum forming or pressure forming, called free forming or free blowing, is used with acrylic sheeting to produce parts that require superior optical quality (e.g., aircraft canopies). In this process the periphery is defined mechanically by clamping, but no mold is used, and the depth of draw or height is governed only by the vacuum or compressed air applied.

1.11.3 Mechanical Forming

Various mechanical techniques have been developed for thermoforming that use neither air pressure nor vacuum. Typical of these is matched mold forming (Figure 1.41). A male mold is mounted on

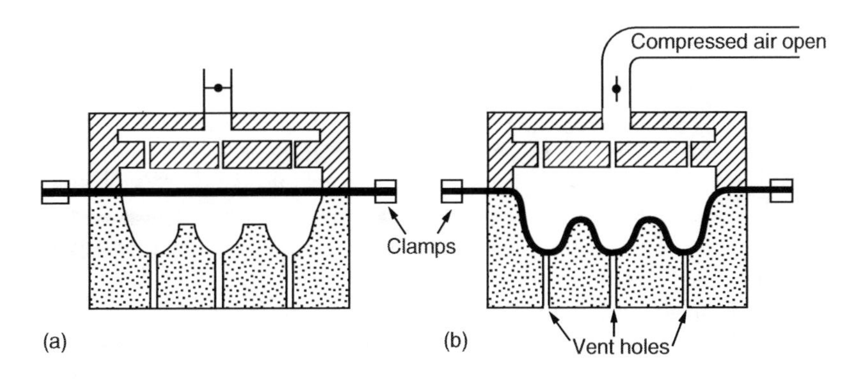

FIGURE 1.40 Pressure forming: (a) heated sheet is clamped over mold cavity; (b) compressed air pressure forces the sheet into the mold.

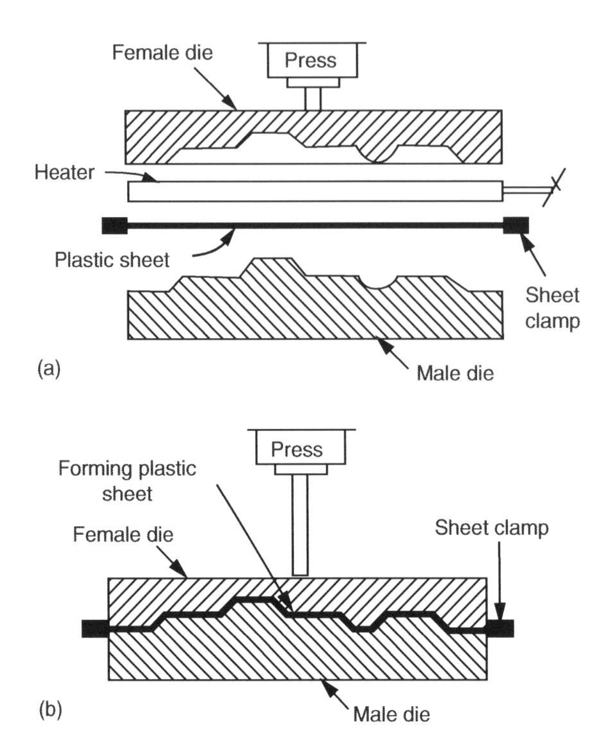

FIGURE 1.41 Matched mold forming: (a) heating; (b) forming.

the top or bottom platen, and a matched female mold is mounted on the other. The plastic sheet, held by a clamping frame, is heated to the proper forming temperature, and the mold is then closed, forcing the plastic to the contours of both the male and the female molds. The molds are held in place until the plastic cools and attains dimensional stability, the latter facilitated by internal cooling of the mold. The matched mold technique affords excellent reproduction of mold detail and dimensional accuracy.

1.12 Casting Processes

There are two basic types of casting used in plastics industry: simple casting and plastisol casting.

1.12.1 Simple Casting

In simple casting, the liquid is simply poured into the mold without applying any force and allowed to solidify. Catalysts that cause the liquid to set are often added. The resin can be a natural liquid or a granular solid liquefied by heat. After the liquid resin is poured into the closed mold, the air bubbles are removed and the resin is allowed to cure either at room temperature or in an oven at low heat. When completely cured, the mold is split apart and the finished casting is removed. In the production of simple shapes such as rods, tubes, etc., usually two-piece metal mold with an entry hole for pouring in the liquid resin is used. For making flat-cast acrylic plexiglass or lucite sheets, two pieces of polished plate glass separated by a gasket with the edge sealed and one corner open are usually used as a mold.

Both thermosets and thermoplastics may be cast. Acrylics, polystyrene, polyesters, phenolics, and epoxies are commonly used for casting.

1.12.2 Plastisol Casting

Plastisol casting, commonly used to manufacture hollow articles, is based on the fact that plastisol in fluid form is solidified as it comes in contact with a heated surface [24]. A plastisol is a suspension of PVC in a liquid plasticizer to produce a fluid mixture that may range in viscosity from a pourable liquid to a heavy paste. This fluid may be sprayed onto a surface, poured into a mold, spread onto a substrate, etc.

The plastisol is converted to a homogeneous solid ("vinyl") product through exposure to heat [e. g., 350°F (176°C)], depending on the resin type and plasticizer type and level. The heat causes the suspended resin to undergo fusion—that is, dissolution in the plasticizer (Figure 1.42)—so that on cooling, a flexible vinyl product is formed with little or no shrinkage. The product possesses all the excellent qualities of vinyl plastics.

Dispersion-grade PVC resins are used in plastisols. These resins are of fine particle size (0.1–2 μm in diameter), as compared to suspension type resins (commonly 75–200 μm in diameter) used in calender and extrusion processing. A plastisol is formed by simply mixing the dispersion-grade resin into the plasticizer with sufficient shearing action to ensure a reasonable dispersion of the resin particles. (PVC plasticizers are usually monomeric phthalate esters, the most important of them being the octyl esters based on 2-ethylhexyl alcohol and isooctyl alcohol, namely dioctyl phthalate and diisooctyl phthalate, respectively.) The ease with which virtually all plastisol resins mix with plasticizer to form a smooth stable dispersion/paste is due to the fine particle size and the emulsifier coating on the resin particles. (The emulsifier coating aids the wetting of each particle by the plasticizer phase.)

The liquid nature of the plastisol system is the key to its ready application. The plastisol may be spread onto a cloth, paper, or metal substrate, or otherwise cast or slushed into a mold. After coating or molding, heat is applied, which causes the PVC resin particles to dissolve in the plasticizer and form a cohesive mass, which is, in effect, a solid solution of polymer in plasticizer.

The various changes a plastisol system goes through in the transformation from a liquid dispersion to a homogeneous solid are schematically shown in Figure 1.42. At 280°F (138°C) the molecules of plasticizer begin to enter between the polymer units, and fusion beings. If the plastisol were cooled after being brought to this temperature, it would give a cohesive mass with a minimum of physical strength. Full fusion occurs and full strength is accomplished when the plastisol is brought to approximately 325°F (163°C) before cooling. The optimum fusion temperature, however, depends on resin type and plasticizer type.

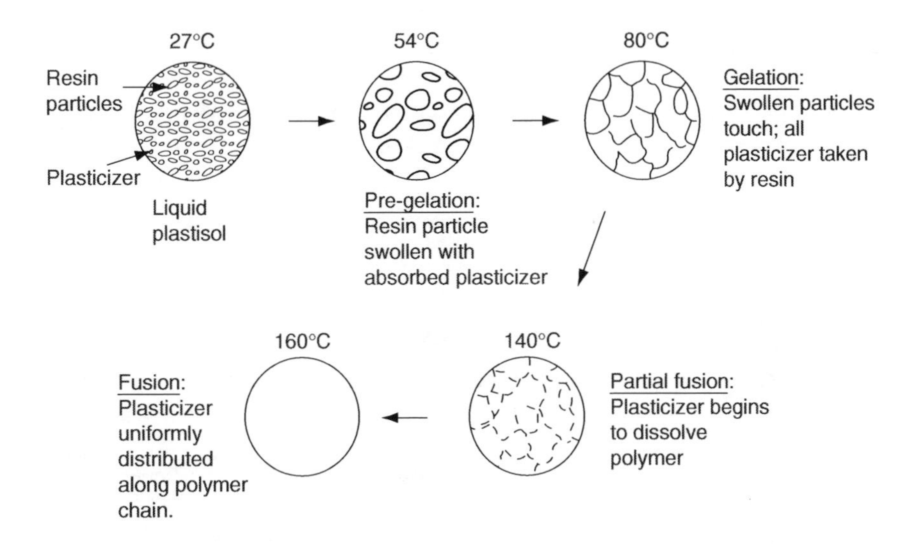

FIGURE 1.42 Various changes in a plastisol system in the transformation from a liquid dispersion to a homogenous solid.

For coating applications it is common practice to add solvent (diluent) to a plastisol to bring down viscosity. This mixture is referred to as organosol. It may be applied by various coating methods to form a film on a substrate and then is heated to bring about fusion, as in the case of plastisol.

Unlike coating applications, there are some applications where it is desirable to have an infinite viscosity at low shear stress. For such applications, a plastisol can be gelled by adding a metallic soap (such as aluminum stearate) or finely divided filler as a gelling agent to produce a plastigel. A plastigel can be cold molded, placed on a pan, and heated to fusion without flow. The whole operation is like baking cookies.

A rigidsol is a plastisol of such formulation that it becomes a rigid, rather than a flexible, solid when fused. A very rigid product can be obtained when the plasticizer is polymerized during or right after fusion. For example, a rigidsol can be made from 100 parts of PVC resin, 100 parts of triethylene glycol dimethacrylate (network forming plasticizer) and 1 part of di-*tert*-butyl peroxide (initiator). This mixture has a viscosity of only 3 poises compared with 25 poises for phthalate-based plastisol. However, after being heated for 10 min at 350°F (176°C), the resin solvates and the plasticizer polymerizes to a network structure, forming a hard, rigid glassy solid with a flexural modulus of over 2.5×10^5 psi (1.76×10^4 kg/cm^2) at room temperature.

Three important variations of the plastisol casting, are dip casting, slush casting, and rotational casting.

1.12.2.1 Dip Casting

A heated mold is dipped into liquid plastisol (Figure 1.43a) and then drawn at a given rate. The solidified plastisol (with mold) is then cured in an oven at 350°F–400°F (176°C–204°C). After it cools, the plastic is stripped from the mold. Items with intricate shapes such as transparent ladies' overshoes, flexible gloves, etc., can be made by this process.

The dipping process is also used for coating metal objects with vinyl plastic. For example, wire dish drainers, coat hangers, and other industrial and household metal items can be coated with a thick layer of flexible vinyl plastic by simply dipping in plastisol and applying fusion.

1.12.2.2 Slush Casting

Slush casting is similar to slip casting (drain) of ceramics. The liquid is poured into a preheated hollow metal mold, which has the shape of the outside of the object to be made (Figure 1.43b). The plastisol in immediate contact with the walls of the hot mold solidifies. The thickness of the cast is governed by the time of stay in the mold. After the desired time of casting is finished, the excess liquid is poured out and the solidified plastisol with the mold is kept in an oven at 350°F–400°F (176°C–204°C). The mold is then opened to remove the plastic part, which now bears on its outer side the pattern of the inner side of the

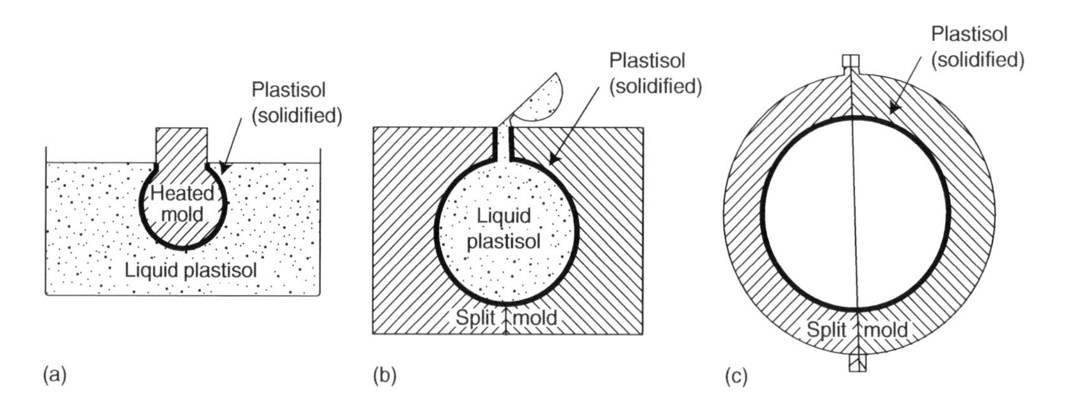

FIGURE 1.43 Plastisol casting processes: (a) dip casting; (b) slush casting; (c) rotational casting.

metal mold. Slush molding is used for hollow, open articles. Squeezable dolls or parts of dolls and boot socks are molded this way.

1.12.2.3 Rotational Casting

In rotational casting a predetermined amount of liquid plastisol is placed in a heated, closed, two-piece mold. The liquid is uniformly distributed against the walls of the mold in a thin uniform layer (Figure 1.43c) by rotating the mold in two planes. The solidified plastisol in the mold is cured in an oven; the mold is then opened, and the part is removed. The method is used to make completely enclosed hollow objects. Doll parts, plastic fruits, squeeze bulbs, toilet floats, etc. can be made by rotational casting of plastisols.

1.13 Reinforcing Processes

An RP consists of a polymeric resin strengthened by the properties of a reinforcing material [25]. Reinforced plastics occupy a special place in the industry. They are at one and the same time both unique materials into themselves and part and parcel of virtually every other segment of the plastics industry.

Reinforced plastics are composites in which a resin is combined with a reinforcing agent to improve one or more properties of the resin matrix. The resin may be either thermosetting or thermoplastic. Typical thermosetting resins used in RPs include unsaturated polyester, epoxy, phenolic, melamine, silicone, alkyd, and diallyl phthalate. In the field of reinforced thermoplastics (RTPs), virtually every type of thermoplastic material can be, and has been, reinforced and commercially molded. The more popular grades include nylon, polystyrene, polycarbonate, polypropylene, polyethylene, acetal, PVC, ABS, styrene-acrylonitrile, polysulfone, polyphenylene sulfide, and thermoplastic polyesters.

The reinforcement used in RP is a strong inert material bound into the plastic to improve its strength, stiffness, or impact resistance. The reinforcing agent can be fibrous, powdered, spherical, crystalline, or whisker, and made of organic, metallic, or ceramic material. Fibrous reinforcements are usually glass, although asbestos, sisal, cotton and high-performance fibers (discussed later) are also used. To be structurally effective, there must be a strong adhesive bond between the resin and the reinforcement. Most reinforcements are thus treated with sizes or finishes to provide maximum adhesion by the resins.

Although by definition, all RPs are composites (i.e., combinations of two materials—resin and reinforcement—that act synergistically to form a new third material RP with different properties than the original components) the term *advanced composites*, or *high-strength composites*, has taken on a special meaning. The term is applied to stiffer, higher modulus combinations involving exotic reinforcements such as graphite, boron, or other high-modulus fibers like aromatic polyamide fibers (Nomex and Kevlar) and extended-chain polyethylene fibers (Spectra ECPE), and resins like epoxy or some of the newer high heat-resistant plastics—polyamides, polyamideimide, polyquinoxalines, and polyphenylquinoxalines. Prime outlets for these materials are in the aerospace and aircraft industries.

1.13.1 Molding Methods

As mentioned above, either thermosetting and thermoplastic resin can be used as the matrix component in RPs. Today, the RTP have become an accepted part of the RPs business, although smaller than the reinforced thermosets. RTP are generally made available to the processor in the form of injection molding pellets (into which glass has been compounded) or concentrates (also a pellet but containing a much higher percentage of reinforcement, and designed to be mixed with nonreinforced pellets). It is also possible for the processor to do his own compounding of chopped glass and thermoplastic powder in injection molding.

In rotational molding and casting techniques, the processor usually adds his own reinforcement. For thermoforming, glass-RTP laminates are sold commercially. Structural foam molding of glass-RTPs and reinforcing molded urethane foams appeared as later developments in the field of RTPs.

Among thermosetting resins, unsaturated polyesters are by far the most widely used in RPs, largely because of their generally good properties, relatively easy handling, and relatively low cost. For special uses, however, other types are significant: epoxies for higher strength, phenolics for greater heat resistance, and silicones for their electrical properties and heat resistance. All these resins must be used in conjunction with a system of catalysts or curing agents in molding thermoset composites. The type and amount strongly affect the properties, working life, and molding characteristics of the resin. The polyester and epoxies are most often mixed with a catalyst just prior to molding. The most widely used catalyst for polyesters is benzoyl peroxide. Where heat is not available for curing, special catalyst-promoter systems can be used. With epoxies, an amine curing agent that reacts with the resin is most often used. However, there are many other types to choose from (see Chapter 1 of *Industrial Polymers, Specialty Polymers, and Their Applications*).

The polyester, epoxy, and thermosetting acrylic resins are usually thick liquids that become hard when cured. For this reason, they are most often combined with the reinforcement, by the molder, by dipping or pouring. There are available, however, preimpregnated reinforcements (prepregs) for the molder who wants to keep the operations as simple as possible.

Several methods are employed to make RPs. Although each method has the characteristics of either molding or casting, the process may be described as (1) hand lay-up. (2) spray-up, (3) matched molding, (4) vacuum-bag molding, (5) pressure-bag molding, (6) continuous pultrusions, (7) filament winding, and (8) prepreg molding.

1.13.1.1 Hand Lay-Up or Contact Molding

A mold is first treated with a release agent (such as wax or silicone-mold release), and a coating of the liquid resin (usually polyester or epoxy) is brushed, rolled, or sprayed on the surface of the mold. Fiberglass cloth or mat is impregnated with resin and placed over the mold. Air bubbles are removed, and the mat or cloth is worked into intimate contact with the mold surface by squeegees, by rollers, or by hand (Figure 1.44a). Additional layers of glass cloth or mat are added, if necessary, to build up the desired thickness. The resin hardens due to curing, as a result of the catalyst or hardener that was added to the resin just prior to its use. Curing occurs at room temperature, though it may be speeded up by heat. Ideally, any trimming should be carried out before the curing is complete, because the material will still be sufficiently soft for knives or shears. After curing, special cutting wheels may be needed for trimming.

Lowest-cost molds such as simple plaster, concrete, or wood are used in this process, since pressures are low and little strength is required. However, dimensional accuracy of the molded part is relatively low and, moreover, maximum strength is not developed in the process because the ratio of resin to filler is relatively high.

The hand lay-up process can be used for fabricating boat hulls, automobile bodies, swimming pools, chemical tanks, ducts, tubes, sheets, and housings, and for building, machinery, and autobody repairs.

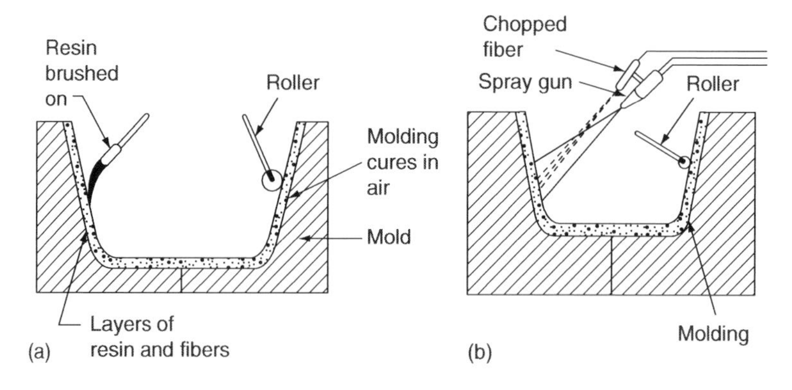

FIGURE 1.44 (a) Basic hand lay-up method. (b) Spray-up technique.

1.13.1.2 Spray-Up

A release agent is first applied on the mold surface, and measured amounts of resin, catalyst, promoter, and reinforcing material are sprayed with a multiheaded spray gun (Figure 1.44b). The spray guns used for this work are different from those used for spraying glazes, enamels, or paints. They usually consist of two or three nozzles, and each nozzle is used to spray a different material. One type, for example, sprays resin and promoter from one nozzle, resin and catalyst from another, and chopped glass fibers from a third. The spray is directed on the mold to build up a uniform layer of desired thickness on the mold surface. The resin sets rapidly only when both catalyst and promoter are present. This method is particularly suitable for large bodies, tank linings, pools, roofs, etc.

1.13.1.3 Matched Metal Molding

Matched metal molding is used when the manufacture of articles of close tolerances and a high rate of production are required. Possible methods are perform molding, sheet molding, and dough molding. In *preform molding* the reinforcing material in mat or fiber form is preformed to the approximate shape and placed on one-half of the mold, which was coated previously with a release agent. The resin is then added to the preform, the second half of the mold (also coated previously with a release agent) is placed on the first half, and the two halves of the mold are then pressed together and heated (Figure 1.45). The resin flows, impregnates the preform, and becomes hard. The cured part is removed by opening the mold. Because pressures of up to 200 psi (14 kg/cm^2) can be exerted upon the material to be molded, a higher ratio of glass to resin may be used, resulting in a stronger product. The cure time in the mold depends on the temperature, varying typically from 10 min at 175°F (80°C) to only 1 min at 300°F (150°C). The cure cycle can thus be very short, and a high production rate is possible.

The molding of sheet-molding compounds (SMC) and dough-molding compounds (DMC) is done "dry"—i.e., it is not necessary to pour on resins. SMC, also called *prepreg*, is basically a polyester resin mixture (containing catalyst and pigment) reinforced with chopped strand mat or chopped roving and formed into a pliable sheet that can be handled easily, cut to shape, and placed between the halves of the heated mold. The application of pressure then forces the sheet to take up the contours of the mold.

DMC is a doughlike mixture of chopped strands with resin, catalyst, and pigment. The charge of dough, also called premix, may be placed in the lower half of the heated mold, although it is generally wise to preform it to the approximate shape of the cavity. When the mold is closed and pressure is applied, DMC flows readily to all sections of the cavity. Curing generally takes a couple of minutes for mold temperatures from 250 to 320°F (120°C–160°C). This method is used for the production of switch gear, trays, housings, and structural and functional components.

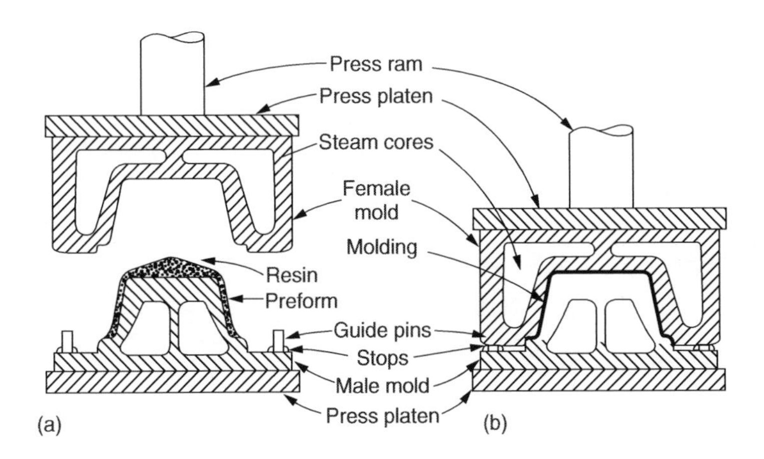

FIGURE 1.45 Matched metal molding: (a) before closing of die; (b) after closing of die.

1.13.1.4 Vacuum-Bag Molding

In vacuum-bag molding the reinforcement and the resin mixed with catalyst are placed in a mold, as in the hand layup method, and an airtight flexible bag (frequently rubber) is place over it. As air is exhausted from the bag, atmospheric air forces the bag against the mold (Figure 1.46). The resin and reinforcement mix now takes the contours of the mold. If the bag is placed in an autoclave or pressure chamber, higher pressure can be obtained on the surface. After the resin hardens, the vacuum is destroyed, the bag opened and removed, and the molded part obtained. The technique has been used to make automobile body, aircraft component, and prototype molds.

1.13.1.5 Pressure-Bag Molding

In pressure-bag molding the reinforcement and the resin mixed with catalyst are placed in a mold, and a flexible bag is placed over the wet lay-up after a separating sheet (such as cellophane) is laid down. The bag is then inflated with an air pressure of 20–50 psi (1.4–3.5 kg/cm^2). The resin and reinforcement follow the contours of the mold (Figure 1.47). After the part is hardened, the bag is deflated and the part is removed. The technique has been used to make radomes, small cases, and helmets.

1.13.1.6 Continuous Pultrusion

Continuous strands, in the form of roving or other forms of reinforcement, are impregnated with liquid resin bath and pulled through a long, heated steel die which shapes the product and controls the resin content. The final cure is effected in an oven through which the stock is drawn by a suitable pulling device. The method produces shapes with high unidirectional strength (e.g., I-beams, rods, and shafts). Polyesters account for 90% of pultrusion resin, epoxies for the balance.

1.13.1.7 Filament Winding

In the filament-winding method, continuous strands of glass fiber are used in such a way as to achieve maximum utilization of the fiber strength. In a typical process, rovings or single strands are fed from a reel through a bath of resin and wound on a suitably designed rotating mandrel. Arranging for the resin impregnated fibers to tranverse the mandrel at a controlled and predetermined (programmed) manner (Figure 1.48) makes it possible to lay down the fibers in any desired fashion to give maximum strengths in the direction required. When the right number of layers have been applied, curing is done at room temperature or in an oven. For open-ended structures, such as cylinders or conical shapes, mandrel design is comparatively simple, either cored or solid steel or aluminum being ordinarily used for the purpose. For structures with integrally wound end closures, such as pressure vessels, careful consideration must be given to mandrel design and selection of mandrel material. A sand-poly(vinyl alcohol) combination, which disintegrates readily in hot water, is an excellent choice for diameters up to 5 ft (1.5 m). Thus, a mandrel made of sand with water-soluble poly(vinyl alcohol) as a binder can be

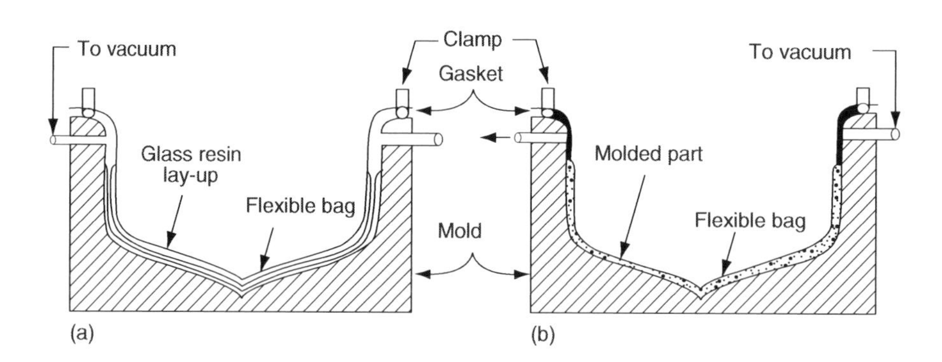

FIGURE 1.46 Vacuum-bag molding: (a) before vacuum applied; (b) after vacuum applied.

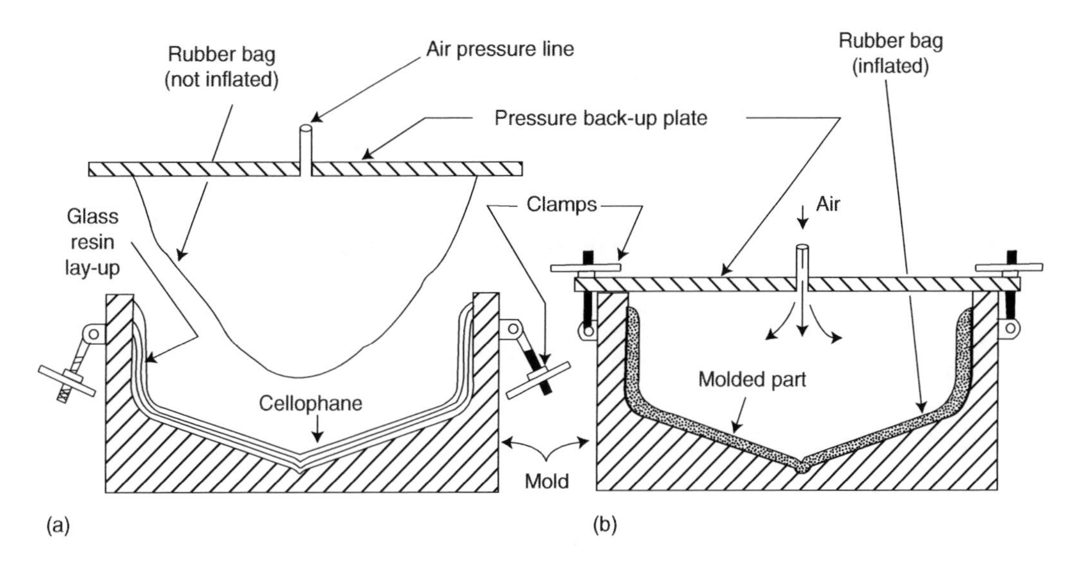

FIGURE 1.47 Pressure-bag molding: (a) during lay-up; (b) during curing.

decomposed with water to recover the filament-wound part. Other mandrel materials include low-melting alloys, eutectic salts, soluble plasters, frangible or breakout plasters, and inflatables.

Because of high glass content, filament-wound parts have the highest strength-to-weight ratio of any reinforced thermoset. The process is thus highly suited to pressure vessels where reinforcement in the highly stressed hoop direction is important. Pipe installation, storage tanks, large rocket motor cases, interstage shrouds, high-pressure gas bottles, etc., are some of the products made of filament winding. The main limitation on the process is that it can only be used for fabricating objects which have some degree of symmetry about a central axis.

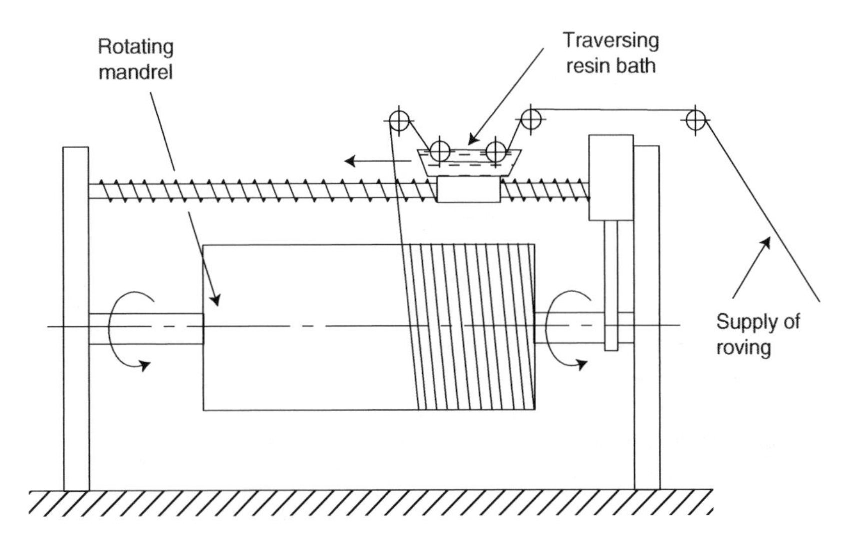

FIGURE 1.48 Sketch of filament winding.

1.13.1.8 Prepreg Molding

Optimal strength and stiffness of continuous fiber–reinforced polymeric composites is obtained through controlled orientation of the continuous fibers. One means to achieve this is by prepreg molding. In this process, unidirectionally oriented layers of fibers are pre-impregnated with the matrix resin and cured to an intermediate stage of polymerization (B-stage). When desired, this pre-impregnated composite precursor, called a prepreg, can be laid up in the required directions in a mold for quick conversion into end components through the use of hot curing techniques. Prepregs can thus be described as pre-engineered laminating materials for the manufacture of fiber–reinforced composites with controlled orientation of fibers.

For the designer, a precisely controlled ply of prepreg represents a building block, with well-defined mechanical properties from which a structure can be developed with confidence. For the fabricator, on the other hand, prepregs provide a single, easy-to-handle component that can be applied immediately to the lay-up of the part to be manufactured, be it aircraft wing skin or fishing rod tube. The prepreg has the desired handleability already built in to suit the lay-up and curing process being utilized, thus improving efficiency and consistency.

Prepregs have been used since the late 1940s, but they have only achieved wide prominence and recognition since the development of the higher performance reinforcing fibers, carbon, and kevlar. The quantum leap in properties provided by these new fibers generated a strong development effort by prepreg manufacturers and there is no doubt that new developments made in this area have been as significant, if not as evident, as that of the introduction of the new fibers.

The use of prepreg in the manufacture of a composite components offers several advantages over the conventional wet lay-up formulations:

1. Being a readily formulated material, prepreg minimizes the material's knowledge required by a component manufacturer. The cumbersome process of stocking various resins, hardeners, and reinforcements is avoided.
2. With prepreg a good degree of alignment in the required directions with the correct amount of resin is easily achieved.
3. Prepreg offers a greater design freedom due to simplicity of cutting irregular shapes.
4. Material wastage is virtually eliminated as offcuts of prepregs can be used as random molding compounds.
5. Automated mass-production techniques can be used for prepreg molding, and the quality of molded product is reproducible.
6. Toxic chemical effects on personnel using prepregs are minimized or eliminated.

A flowchart showing the key stages in the fabrication of composite structures from raw materials with an intermediate prepregging step is given in Figure 1.49. It can be seen that there are two basic constituents to prepreg—the reinforcing fiber and the resin system. All of the advanced reinforcing fibers are available in continuous form, generally with a fixed filament diameter. The number of filaments that the supplier arranges into a "bundle" (or yarn) varies widely, and is an important determinant of the ability to weave fabric and make prepregs to a given thickness. In the majority of cases, the yarn is treated with a size to protect it from abrasion during the weaving or prepregging process. Often, the size is chosen such that it is compatible with the intended resin system.

Resin systems have developed into extremely complex multi-ingredient formulations in an effort to ensure the maximum property benefit from the fiber. Normally there are four methods of impregnation: (1) solution dip; (2) solution spray; (3) direct hot-melt coat; and (4) film calendaring.

The solution dip and solution spray impregnation techniques work with a matrix resin dissolved in a volatile carrier. The low viscosity of the resin solution allows good penetration of the reinforcing fiber bundles with resin. In solution dipping, the fiber, in yarn or fabric form, is passed through the resin solution and picks up an amount of solids dependent upon the speed of through-put and the solids level.

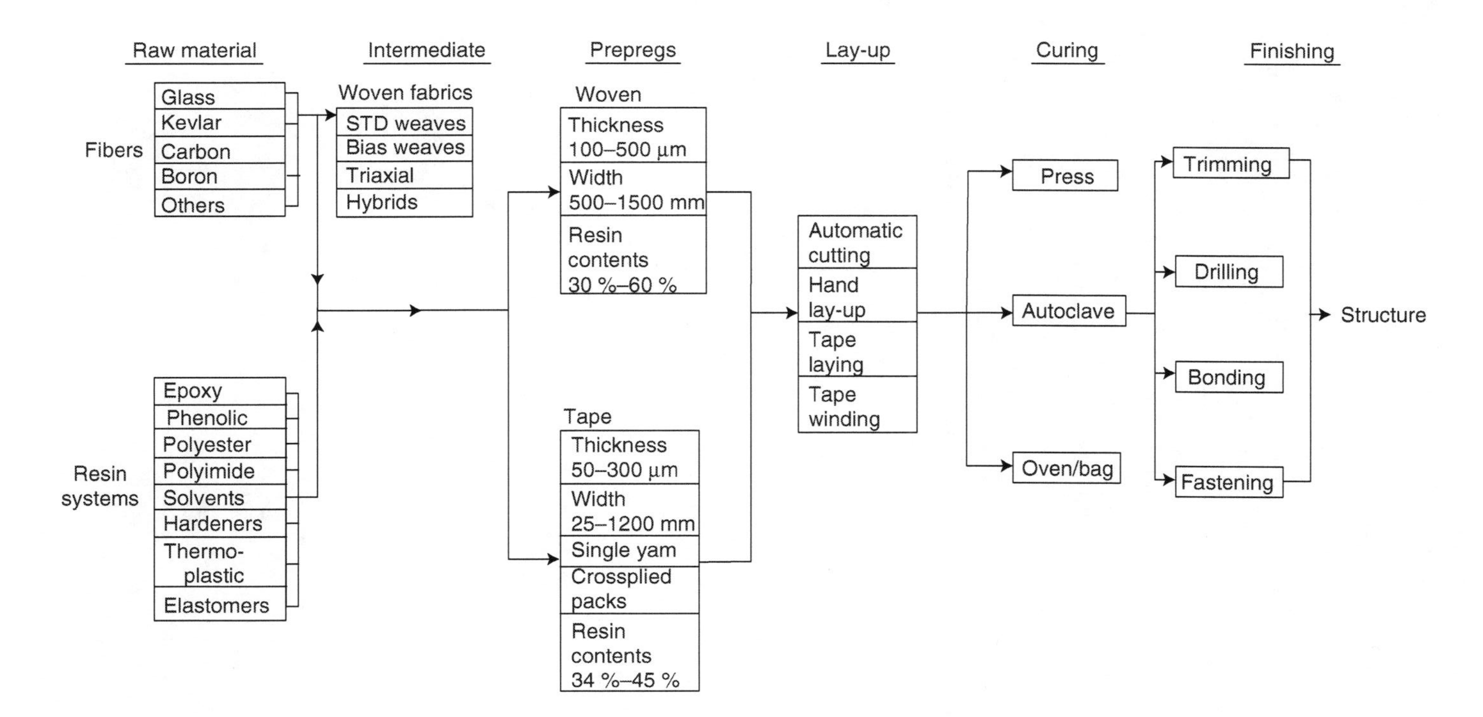

FIGURE 1.49 Flow chart showing key stages in the fabrication of composite structures from raw materials by prepreg molding. (After Lee, W. J., Seferis, J. C., and Bonner, D. C. 1986. *SAMPE Q.*, 17, 2, 58.)

With solution spraying, on the other hand, the required amount of resin formulation is metered directly onto the fiber. In both cases, the impregnated fiber is then put through a heat cycle to remove the solvent and "advance" the chemical reaction in the resin to give the correct degree of tack.

Direct hot melt can be performed in a variety of ways. In one method, the reinforcing fiber web is dipped into a melt resin bath. A doctor blade, scraper bar, or metering roller controls the resin content. Alternately, the melt resin is first applied to a release paper, the thickness of the resin being determined by a doctor blade. The melt resin on the release paper is then brought into contact with a collimated fiber bundle and pressed into it in a heated impregnation zone.

In film calendaring, which is a variation of the above method, the resin formulation is cast into a film from either hot-melt or solution and then stored. Thereafter in a separate process, the reinforcing fiber is sandwiched between two films and calendered so that the film is worked into the fiber.

The decision of which method to use is dependent upon several factors. Holt melt film processes are faster and cheaper, but certain resin formulations cannot be handled in this way, and hence solution methods have to be used. The solution dip method is often preferred for fabrics as the need to squeeze hot-melt and film into the interstices of the fabric can cause distortion of the weave pattern.

In a process based on a biconstituent two impregnation concept, the polymeric matrix is introduced in fibrous form and a comingled tow of polymer and reinforcing fibers are fed into a heated impregnation zone (Figure 1.50). In this method, it may be possible to effect better wetout of the reinforcing fibers with the matrix polymer, especially with high viscosity thermoplastic matrix polymers, through intimate comingling.

The machines necessary to accomplish the above prepregging procedures are many and varied. There are three distinct aspect to quality control: raw material screening, on-line control, and batch testing. All three are obviously important, but the first two are more critical.

Fibers and base resins are supplied against certificates of conformance and often property test certificates. On-line control during the manufacture of prepreg revolves around the correct ratio of fiber to resin. This is done by a traversing Beta-gauge, which scans the dry fiber (either unidirectional or fabric) and then the impregnated fiber and provides a continuous real-time plot of the ratio across the width of the material. This can be linked back to the resin system application point for continuous adjustment.

Batch testing is carried out to verify prepreg properties, such as resin content, volatile level, and flow. The resin "advancement" (chemical reaction) is monitored via a Differential Scanning Calorimeter (DSC) and the formulation consistency by testing the T_g via DSC or Dynamic Mechanical Analyzer (DMA). The laminate properties are also determined. All are documented and quoted on a Release Certificate.

Commercial prepregs are available with different trade names. Fibredux 914 of Ciba–Geigy is a modified epoxy resin preimpregnated into unidirectional fibers of carbon (HM or HT), glass (E type and R type), or aramid (Kevlar 49) producing prepregs that, when cured to form fiber reinforced composite components, exhibit very high strength retention between $-60°C$ and $+180°C$ operational temperatures.

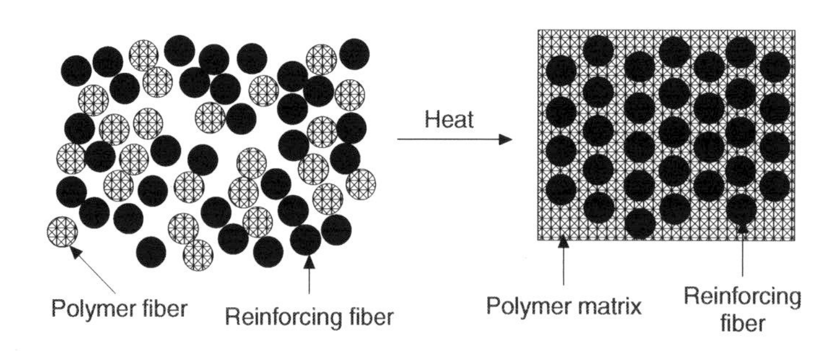

FIGURE 1.50 Biconstituent tow impregnation using a comingled tow of polymer and reinforcing fibers. (After Lee, W. J., Seferis, J. C., and Bonner, D. C. 1986. *SAMPE Q.*, 17(2), 58.)

"Scotchply" brand RPs of Industrial Specialties Division of 3M Company are structural-grade thermosetting molding materials, consisting of unidirectional nonwoven glass fibers embedded in an epoxy resin matrix. The product is available both in prepreg form in widths up to 48 in. (1.22 m) and in flat sheet stock in sizes up to 48 in. (1.22 m) $\times$ 72 in. (1.83 m). The cured product is claimed to possess extraordinary fatigue life, no notch sensitivity, high ultimate strength, and superior corrosion resistance. The range of application of the RP includes vibratory springs, sonar housings, landing gear, picker blades, snowmobile track reinforcement, helicopter blades, seaplane pontoons, missile casing, and archery bow laminate.

1.13.2 Fibrous Reinforcements

Although many types of reinforcements are used with plastics, glass fibers predominate. Fibrous glass reinforcements are available in many forms (described below). Asbestos is used in the form of loose fiber, paper, yarn, felt, and cloth. The two largest uses of asbestos in plastics are with PVC in vinyl asbestos tile and with polyesters and polypropylene.

Most natural and synthetic fibers do not have the strength required for an RPs part. However, when intermediate strengths are satisfactory, they can be used. In this category are nylon, rayon, cotton fabrics, and paper. Sisal fibers have also found use as a low cost reinforcing material in premix molding compounds.

High-modulus graphite and carbon fibers, aramid fibers and ECPE fibers are playing a more and more important role in RPs. Boron filaments, with outstanding tensile strengths, are usually used in the form of prepreg tapes and have been primarily evaluated for the aerospace and aircraft industry.

1.13.2.1 Glass Fibers

A high-alkali "A-glass" and a low-alkali "E-glass" are used as reinforcements for polymer composites, the latter being used most often. Since the modulus of E-glass is 10.5×10^6 psi and the tensile strength upwards of 250,000 psi, it is not surprising that the stiffness and strength of most plastics can be increased by compounding with glass. A more chemical resistant glass, sodium borosilicate (C-glass), and a higher-tensile-strength glass, S-glass, are also available. E-glass is a calcium alumino-silicate, and S-glass is a magnesium aluminosilicate. Fiberglass is available as a collection of parallel filaments (roving), chopped strands, mat, and woven fabric.

Glass filaments are produced by melting a mixture of silica, limestone, and other reactants, depending on the type of glass and forcing the molten product through small holes (bushings). The hot filaments are gathered together and cooled by a water spray. These multiple glass filaments are gathered together into a bundle, called a strand, which is wound up on a coil. Short fibers (staple) are produced by passing a stream of air across the filaments as they emerge from the bushings.

Rovings are rope-like bundles of continuous untwisted strands for use in such processes as preform press molding, filament winding, spray-up, pultrusion, and centrifugal casting. They can also be converted into chopped strand mats or cut into short fibers for molding compounds.

Chopped strands of glass 1/32–1/2 in. in length can be incorporated in thermoset or thermoplastic materials about as easily as the particulate fillers. Each strand may be made up of 204 individual filament whose diameter is $2\text{–}7.5 \times 10^{-4}$ in.

Chopped strands several inches long can be loosely bound as a mat that is porous and in which the strands are randomly oriented in two dimensions. This form is suitable for impregnation by a liquid polymer. After polymerization or cross-linking (curing) under pressure, the composite will comprise a polymer-network matrix in which the individual strands are embedded.

Chopped strand mats provide nondirectional reinforcement (i.e., strengths in many directions, as contrasted to unidirectional forms which are continuous fibers, like roving, that provide strength in one direction). These mats are available in a variety of thicknesses, usually expressed in weight per square foot. In order to hold the fibers together, a resin binder is generally used, the type depending on the resin and molding process. In some cases, the mats are stitched or needled, instead of using the resin binder.

A woven glass fabric (cloth) might be used in place of the mat or in combination with it. In this case there will be a variation in strength with the angle between the axis of the fibers and the direction of stress. Twisted yarns are generally woven into fabrics of varying thickness and with tight or loose weaves, depending upon the application. Most are balanced weaves (i.e., equal amounts of yarn in each direction), although some are unidirectional (more fibers running in one direction). Although costlier, they offer a high degree of strength. Rovings can also be woven into a fabric that is less costly than the woven yearn fabrics, coarser, heavier, and easier to drape.

For optimum adhesion at the interface between the fiber surface (stationary phase) and the resin matrix (continuous phase), the glass fibers must be treated with coupling agents to improve the interfacial adhesion. The pioneer coupling agent (linking agent) was methacrylatochromic chloride (Volan). This has been supplanted by organosilanes, organotitanates, and organozirconates. These coupling agents contain functional groups, one of which is attracted to the fiber surface and the other to the resin (Figure 1.51).

1.13.2.2 Graphite Carbon Fibers

Graphite carbon fibers are the predominant high-strength, high-modulus reinforcing agent used in the fabrication of high-performance polymer composites. In general, the term *graphite fiber* refers to fibers that have been treated above 1,700°C (3,092°F) and have tensile moduli of elasticity of 5×10^5 psi (3,450 MPa) or greater. Carbon fibers are those products that have been processed below 1,700°C (3092°F) and consequently exhibit elastic moduli up to 5×10^5 psi (3,450 MPa) [25]. A further distinction is that the carbon content of carbon fibers is 80%–95%; and that of graphite, above 99%. However, the industry has universally adopted the term "graphite." It will therefore be used to describe both product forms in this section.

Graphite fibers were first utilized by Thomas Edison in 1880 for his incandescent lamps. The filaments were generated by the carbonization of bamboo in the absence of air. When tungsten filaments replaced the graphite in lamps, interest in graphite materials waned until the mid 1950s when rayon-based graphite fibers were created. These products exhibited relatively high tensile strengths of about 4×10^5 psi (2,760 MPa) and were designed for rocket/missile ablative component applications.

A significant event that led to the development of today's graphite industry was the utilization of PAN as a graphite precursor material by Tsunoda in 1960 [27]. Subsequent work led to continued improvement of PAN-based graphite fiber properties by numerous researchers. These development focused on stretching the PAN precursor to obtain a high degree of molecular orientation of the polymer molecules followed by stabilizing it under tensile load, carbonization, and graphitization. PAN-based graphite fibers are now available with tensile moduli of up to 1.2×10^6 psi (8,280 MPa) and tensile strengths above 8×10^5 psi (5,516 MPa).

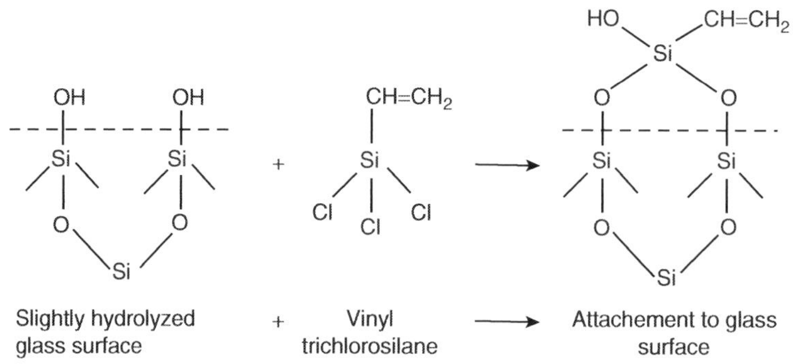

Slightly hydrolyzed glass surface + Vinyl trichlorosilane → Attachement to glass surface

FIGURE 1.51 Mechanism of functioning of a glass surface finish.

Pitch was first identified as a graphite precursor by Otani in 1965 [28]. These fibers are made by melt spinning a low-cost isotropic molten (petroleum) pitch and then oxidizing the filaments as they are spun. This step is followed by carbonization at 1,000°C (1832°F) in an inert atmosphere. Process modifications to improve the fiber properties evolved through the 1970s until pitch-based (including mesophase liquid crystal pitch) graphite fibers with tensile strengths up to 3.75×10^5 psi (2,590 MPa) and tensile moduli to 1.2×10^6 psi (8,300 MPa) were achievable.

1.13.2.3 Fiber Manufacture

The pyrolysis of organic fibers used as graphite precursors is a multistage process. The three principal graphite precursors are PAN, pitch, and rayon, with PAN as the predominant product.

The commercial production of PAN precursor fiber is based on either dry or wet spinning technology. In both instances, the polymer is dissolved in either an organic or inorganic solvent at a concentration of 5–10% by weight. The fiber is formed by extruding the polymer solution through spinneret holes into hot gas environment (dry spinning) or into a coagulating solvent (wet spinning). The wet spinning is more popular as it produces fibers with round cross-section, whereas the dry spinning results in fiber with a dog bone cross-section.

The wet-spun precursor making process [29] includes three basic steps: polymerization, spinning, and after treatments (Figure 1.52). Acrylonitrile monomer and other comonomers (methyl acrylate or vinyl acetate) are polymerized to form a PAN copolymer. The reactor effluent solution, called "dope," is purified, the unreacted monomers removed, and the solid contaminants filtered off. The spinning process next extrudes the purified dope through holes in spinnerettes into a coagulating solution. The spun gel fiber then passes through a series of after-treatments such as stretching, oiling, and drying. The product is the PAN precursor.

In order to produce high-strength, high-modulus graphite fibers from the PAN precursor, it is essential to produce preferred molecular orientation parallel to the fiber axis and then "stabilize" the fiber against relaxation phenomena and chain scission reactions that may occur in subsequent carbonization steps. A typical step-by-step PAN-based graphite manufacturing process begins with the aforementioned precursor stabilization which is followed by carbonization, graphitization, surface treatment, and sizing, as shown schematically in Figure 1.53.

The stabilization of the PAN precursor involves "preoxidation" by heating the fiber in an air oven at 200°C–300°C (392°F–5722°F) for approximately one hour while controlling the shrinkage/tension of the fiber so that the PAN polymer is converted into a thermally infusible aromatic ladder-like structure.

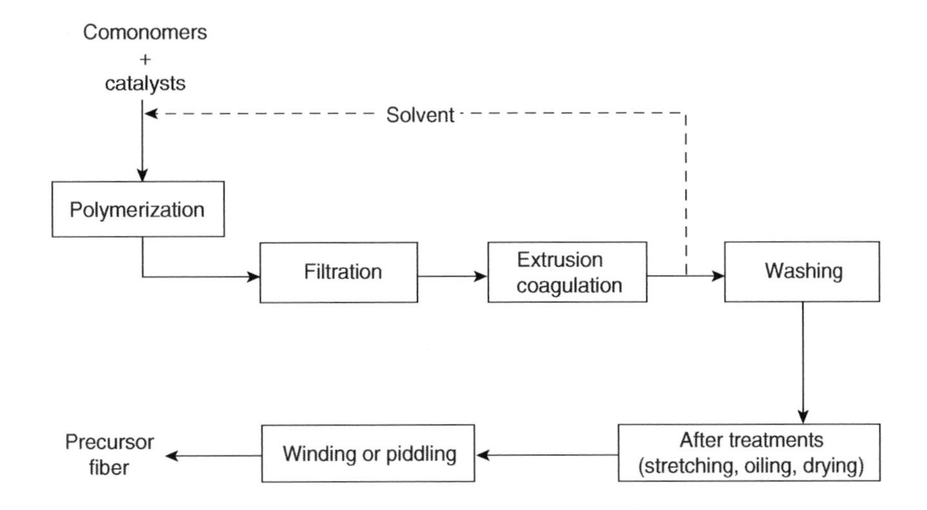

FIGURE 1.52 Typical PAN precursor manufacturing steps based on solution polymerization and wet spinning.

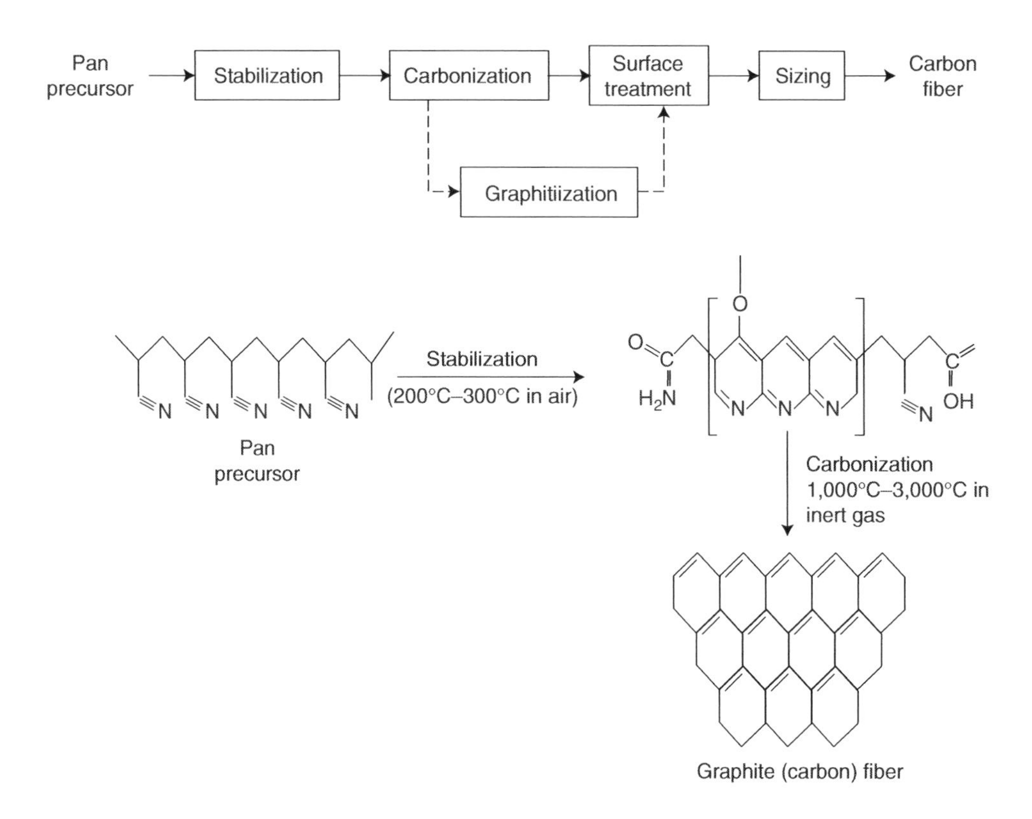

FIGURE 1.53 Schematic of a typical step-by-step PAN-based graphite manufacturing process.

The next step is the process of carbonization, which pyrolyzes the stabilized PAN-based fibers until they are transformed into graphite (carbon) fibers. The carbonization treatment is done in an inert atmosphere (generally nitrogen) at temperatures greater than 1,200°C (2,192°F). This step removes hydrogen, oxygen, and nitrogen atoms from the ladderlike polymers whose aromatic rings then collapse into a graphitelike polycrystalline structure. It is during this stage that high-mechanical-property characteristics of graphite fibers are developed. The development of these properties is directly related to the formation and orientation of graphitelike fibers or ribbons within each individual fiber.

Graphitization performed at temperatures above 1,800°C (3,272°F) is an optional treatment. Its purpose is to improve the tensile modulus of elasticity of the fiber by improving the crystalline structure and orientation of graphitelike crystallites within each individual fiber. The higher heat-treatment temperature used in graphitization also results in a higher carbon content of that fiber. The final step in the process of producing graphite or carbon fiber is surface treatment and sizing prior to bobbin winding the continuous filaments. The surface treatment is essentially an oxidation of the fiber surface to promote wettability and adhesion with the matrix resin in the composite. Sizing improves handleability and wettability of the fiber with the matrix resin. Typical sizing agents are poly(vinyl alcohol), epoxy, and polyimide.

Pitch-based graphite fibers are produced by two processes. The precursor of one of these processes is a low-softening-point isotropic pitch and the process scheme includes the following steps: (1) melt-spin isotropic pitch; (2) thermoset at relatively low temperatures for long periods of time; (3) carbonize in an inert atmosphere at 1,000°C (1,832°F); (4) stress graphitize at high temperatures 3,000°C (5,432°F). The high-performance fibers produced in this manner are relatively expensive because of the very long thermosetting time required and the need for high-temperature stress graphitization.

The commercially more significant process for making pitch-based fibers is the mesophase process, which involves the following steps: (1) heat treat in an inert atmosphere at 400–450°C (752–842°F) for an extended period of time in order to transform pitch into a liquid-crystalline (mesophase) state; (2) spin the mesophase pitch into fibers; (3) thermoset the fibers at 300°C (572°F) for 2½ h; (4) carbonize the fibers at 1,000°C (1,832°F); and (5) graphitize the fibers at 3,000°C (5,432°F). Since long thermosetting times and stress graphitization treatment are not required, the high-temperature graphite fibers produced by this process are lower in cost.

The process by which rayon precursor is converted to graphite fibers includes four steps: (1) fiber spinning; (2) stabilization at 400°C (752°F) for long periods of time; (3) carbonization at 1,300°C (2,372°F); and (4) stress graphitization at high temperatures 3,000°C (5,432°F). The rayon-based graphite fibers produced in this manner tend to be relatively expensive because of the very long stabilization times required and the need for stress graphitization at high temperatures.

1.13.2.4 Properties and Applications

The excellent properties of graphite are directly attributable to the highly anisotropic nature of the graphite crystal. The standard-grade PAN-based graphite fibers, which make up the largest part of both the commercial and aerospace markets have tensile strengths ranging from 4.5×10^5 to 5.5×10^5 psi (3,100–3,800 MPa) and moduli of approximately 340,000 psi (2,345 MPa). A family of intermediate-modulus/high-strain fibers with tensile strengths up to 7×10^5 psi (4,800 MPa) and modulus above 4×10^5 psi (2,760 MPa) have been developed to meet high-performance aerospace requirements. The high-modulus fibers (both PAN- and pitch-based), used in high stiffness/low-strength applications, such as space hardware, have tension moduli ranging from 5×10^5 to 1.2×10^6 psi (3,450–8,280 MPa) and strain to failures generally greater than or equal to 1%.

Graphite fibers are available to the user in a variety of forms: continuous filament for filament winding, braiding, or pultrusion; chopped fiber for injection or compression molding; impregnated woven fabrics and unidirectional tapes for lamination.

The major markets for advanced graphite fiber composites are aerospace, marine, automotive, industrial equipment, and recreation. Military aerospace applications dominate the market and military consumptions are slated to increase rapidly as programs, which utilize a very high percentage of composites move from development to large-scale production. Graphite fiber usage in space applications is in a large measure linked to Space Station programs and production activities.

Examples of nonaerospace military applications include portable, rapid deployment bridges for the army and propeller shafts for submarines. Fiber usage in the commercial aerospace sector is also growing. Commercial planes such as Boeing 767 and the Airbus A320 utilize two to three times the graphite fiber per plane that is used in older commercial models.

The biggest industrial market potential of graphite fibers is in the automotive sector. The graphite composite usage in this area should increase as lower-cost fibers become available. A major and growing use of chopped graphite fibers in the industrial market is as a reinforcement for thermoplastic-injection-molding compounds. The advantages of such use include greater strength and stiffness, higher creep and fatigue resistance, increased resistance to wear, higher electrical conductivity, and improved thermal stability and conductivity.

1.13.2.5 Aramid Fibers

Aramid fiber is the generic name for aromatic polyamide fibers. As defined by the U.S. Federal Trade Commission, an aramid fiber is a "manufactured fiber in which the fiber forming substance is a long chain synthetic polyamide in which at least 85% of the amide linkages are attached directly to two aromatic rings":

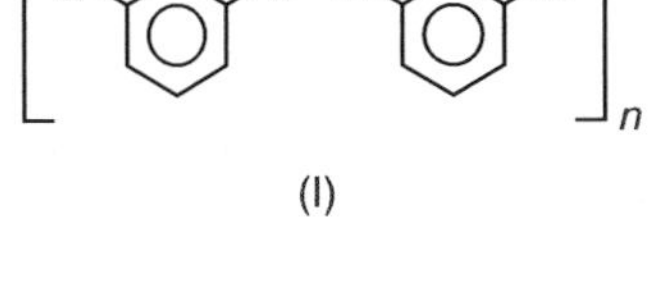

(I)

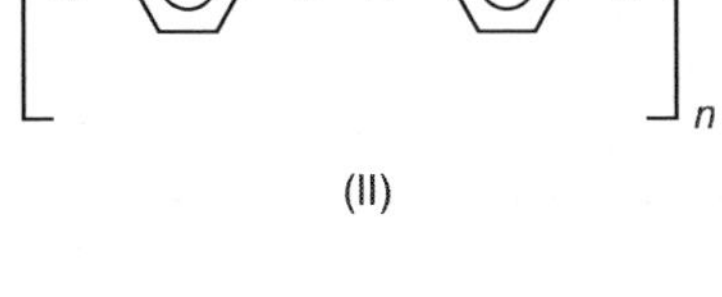

(II)

Among the commercially available aramid fibers are Du Pont's Nomex (I) and Kelvar (II); in fact these trade names are commonly used in lieu of the generic name. Kelvar 49 is a high-modulus aramid fiber and is the most widely used reinforcing aramid fiber. Kevlar 29 has a lower modulus and Kevlar 149 has a higher modulus than Kevlar 49. Aromatic polyamides are described in greater detail in Chapter 1 of *Industrial Polymers, Specialty Polymers, and Their Applications.*

Aramid fibers can be used to advantage to obtain composites having lighter weight, greater stiffness, higher tensile strength, higher impact resistance, and lower notch sensitivity than composites incorporating E-glass or S-glass reinforcement. Weight savings over glass result from the lower specific gravity of aramid fibers, 90.4 lb/in.^{3} (1.45 g/cm^3) versus E-glass, 159.0 lb/in.^{3} (2.55 g/cm^3). Higher stiffnesses are reflected in a Young's modulus up to 19×10^6 psi (1.31×10^5 MPa) for Kevlar 49 and 27×10^6 psi (1.86×10^5 MPa) for Kevlar 149, compared to 10^7 psi (6.9×10^4 MPa) for E-glass and 12×10^6 psi (8.6×10^4 MPa) for S-glass. Aramid composites are more insulating than their glass counterparts, both electrically and thermally, more damped to mechanical and sonic vibrations, and are transparent to radar and sonar. Despite their outstanding mechanical properties, these high-modulus organic fibers have the processability normally associated with conventional textiles. This leads to wide versatility in the form of the reinforcement, e.g., yarns, rovings, woven and knit goods, felts, and papers.

1.13.2.6 Applications

Fabrics woven from Kevlar 49 aramid fiber are often used as composite reinforcement, since fabrics offer biaxial strength and stiffness in a single ply. The mechanical properties of Kevlar 49 aramid are dependent on the fabric construction. The composite properties are functions of the fabric weave and the fiber volume fraction (typically 50%–55% with ply thickness 5–10 mils, depending on fabric construction). In 1987, Du Pont introduced high-modulus Kevlar 149. Compared to Kevlar 49 it has higher performance (47% modulus increase) and lower dielectric properties (65% decrease in moisture regain).

Over the past three decades, Kevlar has gained wide acceptance as a fiber reinforcement for composites in many end uses, such as tennis rackets, golf clubs, shafts, skis, ship masts, and fishing rods. The boating and aircraft industries make extensive use of advanced composites. The advanced composites have allowed innovative designers to move ahead in designing aircraft with unprecedented performance.

The cost of high-modulus aramid fibers is higher than E-glass and equivalent to some grades of S-glass on a unit-weight basis. Price differences versus glass are, however, reduced by about half on a unit volume basis when lower density of the aramids is taken into account.

For many applications, fabrics containing more than one fiber type offer significant advantages. Hybrids of carbon and Kevlar 49 aramid yield greater impact resistance over all-carbon construction and higher compressive strength over all-Kevlar construction. Hybrids of Kevlar 49 and glass offer enhanced properties and lower weight than constructions containing glass as the sole reinforcement and are less expensive than constructions using only Kevlar 49 reinforcement.

1.13.2.7 Extended-Chain Polyethylene Fibers

Extended-Chain Polyethylene Fibers (ECPE) fibers are relatively recent entrants into the high-performance fibers field. Spectra ECPE, the first commercially available ECPE fiber and the first in

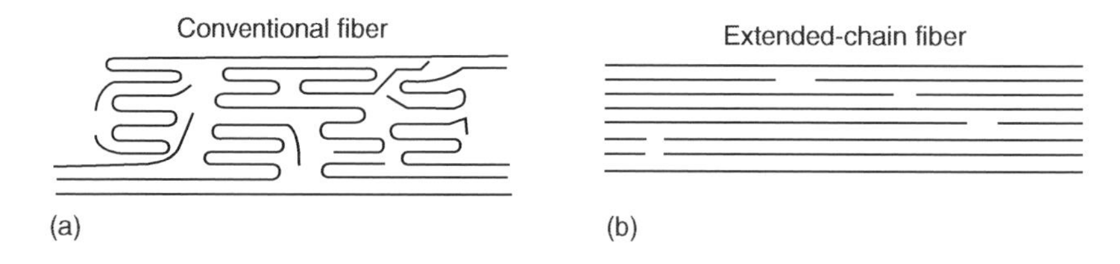

FIGURE 1.54 Fiber morphology of polyethylene. (a) Conventional PE fiber characterized by relatively low molecular weight, moderate orientation, and crystalline regions chain folded. (b) Extended-chain PE fiber characterized by very high molecular weight, very high degree of orientation, and minimum chain folding.

family of extended chain polymers manufactured by Allied-Signal, Inc. was introduced in 1985. ECPE fibers are arguably the highest modulus and highest strength fibers made. These are being utilized as a reinforcement in such applications as ballistic armor, impact shields, and radomes to take advantage of the fiber's unique properties.

Polyethylene is a flexible molecule that normally crystallizes by folding back on itself (see Chapter 1 of *Plastics Fundamentals, Properties, and Testing*). Thus fibers made by conventional melt spinning do not possess outstanding physical properties. ECPE fibers, on the other hand, are made by a process that results in most of the molecules being fully extended and oriented in the fiber direction, producing a dramatic increase in physical properties. Using a simple analogy, the structure of ECPE fibers can be described as that of a bundle of rods, with occasional entangled points that tie the structure together. Conventional PE, by comparison, is comprised of a number of short-length chain folds that do not contribute to material strength (see Figure 1.54). ECPE fibers are, moreover, made from ultrahigh molecular weight polyethylene (UHMWPE) with molecular weight generally 1–5 million that also contributes to superior mechanical properties. Conventional PE fibers, in comparison, have molecular weights in the range 50,000 to several hundred thousand. ECPE fibers exhibit a very high degree of crystalline orientation (95%–99%) and crystalline content (60%–85%).

High-modulus PE fibers can be produced by melt extrusion and solution spinning. The melt extrusion process leads to a fiber with high modulus but relatively low strength and high creep whereas solution spinning in which very high-molecular-weight PE is utilized yields a fiber with both high modulus and high strength. The solution spinning process for a generalized ECPE fiber starts with the dissolution of polyethylene of approximately 1–5 million molecular weight in a suitable solvent. This serves to disentangle the polymer chains, a key step in achieving an extended chain polymer structure. The solution must be fairly dilute to facilitate this process, but viscous enough to be spun using conventional melt spinning equipment. The cooling of the extrudate lends to the formation of a fiber that can be continuously dried to remove solvent or later extracted by an appropriate solvent. The fibers are generally postdrawn prior to final packaging.

The solution spinning process is highly flexible and can provide an almost infinite number of process and product variations of ECPE fibers. Fiber strengths of $(3.75\text{–}5.60) \times 10^5$ psi (2,890–3,860 MPa) and tensile moduli of $(15-30) \times 10^6$ psi $[(103-207) \times 10^3$ MPa$]$ have been achieved. The properties are similar to other high-performance fibers; however, because the density of PE is approximately two-thirds that of high modulus aramid and half that of high-modulus carbon fiber, ECPE fibers possess extraordinarily high specific strengths and specific moduli. Figure 1.54 compares the specific strength versus specific modulus for currently available fibers.

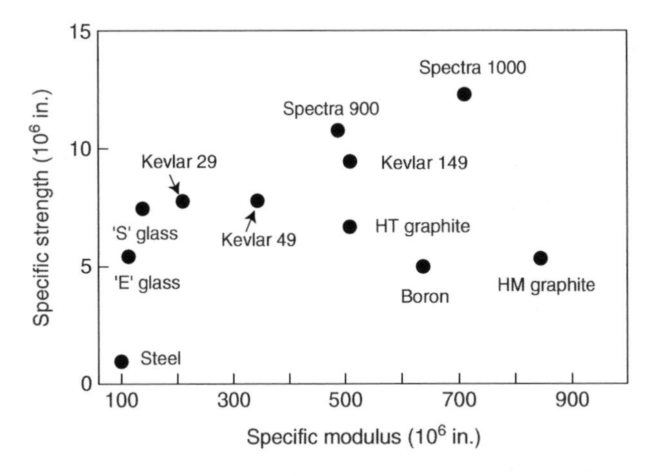

FIGURE 1.55 Comparative properties of various reinforcing fibers.

Traditional binders and wetting agents are ineffective in improving resin adhesion to polyethylene. For ECPE fibers, this characteristic is actually advantageous in specific areas. For instance, ballistic performance is inversely related to the degree of adhesion between the fiber and the resin matrix. However, for applications requiring higher levels of adhesion and wetout, it has been shown that by submitting ECPE fiber to specific surface treatments, such as corona discharge or plasma treatments, the adhesion of the fiber to various resins can be dramatically increased.

The chief application areas being explored and commercialized for ECPE fibers are divided between traditional fiber applications and high-tech composite applications. The former include sailcloth, marine ropes, cables, sewing thread, nettings, and protective clothing. The latter includes impact shields, ballistics, radomes, medical implants, sports equipment, pressure vessels, and boat hulls.

ECPE fibers (such as Spectra 1,000) are well suited for high-performance yachting sails, offering, in addition, resistance to sea water and to typical cleaning solutions used in the sailing industry, such as bleach. The major sport equipment applications to date have been canoes, kayaks, and snow and water skis. Numerous other sport applications are under development.

The high-strength, lightweight, low-moisture absorption and excellent abrasion resistance of ECPE make it a natural candidate for marine rope. In marine rope applications, load, cycling, and abrasion resistance are critical. Thus a 12-strand ECPE braid, for example, is reported to withstand about eight times the number of cycles that cause failure in 12-strand aramid braid.

Specially toughened and dimensionally stabilized ECPE yarn has been used in a revolutionary new line of cut-resistant products. ECPE fibers are being used to produce cut-resistant gloves, arm guards, and chaps in such industries as meat packing, commercial fishing, and poultry processing and in sheet metal work, glass cutting, and power tool use.

ECPE's high strength and modulus and low specific gravity offer higher ballistic protection at lower density per area than is possible with currently used materials. The significant applications include flexible and rigid armor. Flexible armor is manufactured by joining multiple layers of fabric into the desired shape, the ballistic resistance being determined by the style of the fabric and the number of layers. Traditional rigid armor can be made by utilizing woven ECPE fiber in either thermoset or thermoplastic materials. Ballistics are currently the dominant market segment. Products include helmets, helicopter seats, automotive and aircraft armor, armor radomes, and other industrial structures.

The radome (radome protective domes) market is also important for ECPE fibers. ECPE composite systems act as a shield that is virtually transparent to microwave signals, even in high-frequency regimes.

1.14 Reaction Injection Molding

A new type of injection molding called Reaction Injection Molding (RIM) has become important for fabricating thermosetting polymers [30,31]. RIM differs from the conventional injection molding in that the finished product is made directly from monomers or low-molecular-weight polymeric precursors (liquid reactants), which are rapidly mixed and injected into the mold even as the polymerization reaction is taking place. Thus, synthesis of polymer prior to molding is eliminated, and the energy requirements for handling of monomers are much less than those for viscous polymers.

For RIM to be successful, the monomers or liquid reactants must be fast reacting, and the reaction rates must be carefully synchronized with the molding process. Thus the polymers most commonly processed by RIM are polyurethanes and nylons though epoxies, and certain other polymers such as polycyclopentadiene have been processed by RIM. The process uses equipment that meters reactants to an accuracy of 1%, mixes them by high-pressure impingement, and dispenses the mixture into a closed mold. The mold is, in fact, a chemical reactor. The reaction in RIM takes place in a completely filled mold cavity. Reinforcing fillers are sometimes injected into the mold along with the reactants, a process called reinforced reaction injection molding (RRIM). In the mold the functional groups of the liquid

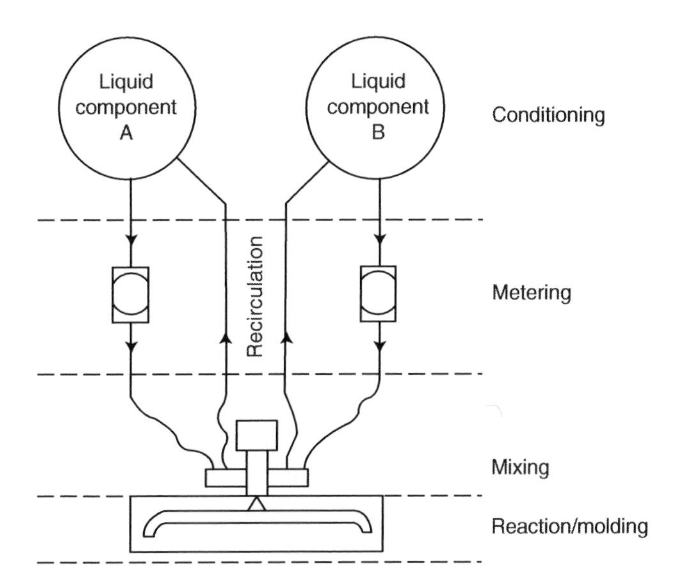

FIGURE 1.56 Schematic of RIM process.

reactants react to form chemical linkages, producing solid polymers, which comprise polymeric chains or networks depending on the starting materials. The temperature of the mold plays a vital role in the polymerization of reactants.

To produce a molded part by the RIM process requires precise but realistic process control. Figure 1.56 shows a simplified schematic of the RIM process. The important elements of the process are conditioning, metering, mixing, and molding. All the liquid reactants require precise temperature control. The flow property (viscosity) of the liquid reactants usually varies with temperature as does the density. For accurate metering the temperature must be controlled within very narrow limits. This is usually accomplished by recirculating reactants from conditioning or storage tanks designed to maintain raw material temperatures specified by the system supplier. These conditions are normally quite moderate (30°C–38°C) for polyurethanes.

For some polymerization systems, such as nylon which is processed at a high temperature, the machines are designed with heated lines and temperature control devices for pumps, mixers, and other components. The molds are also designed to control temperatures as the reaction characteristics of the RIM process are exothermic. A higher temperature increases the reaction rate, which results in an decrease of cycle time. For polyurethane, however, speeding the reaction rate by operating the mold at elevated temperatures is to be avoided, as this changes the types of linkages produced (see "Polyurethanes" in Chapter 1 of *Industrial Polymers, Specialty Polymers, and Their Applications*).

Polyurethane RIM systems have been commercial in the United States for about 50 years and a bit longer in Europe. It is still a rapidly growing field of technology. The automotive industries in the United States account for most of the commercial RIM production. A later development for RIM polyurethane, and to a lesser extent RIM nylons, is the application for housings of various instruments and appliances: computer housings, business machine housings, TV and radio cabinets, instrument cases, and similar electronic product enclosures. While elastomeric RIM is most commonly used in these applications, some housings are also molded from RIM structural foam.

Though systems suppliers do not always clearly differentiate between elastomeric and structural RIM, elastomeric RIM is molded in thin cross-section (usually 0.125 in.) at high density while structural foam has an interior foam structure, a density about one-third that of elastomeric RIM, and is molded in thicker cross-section (usually 0.375 in.).

1.14.1 Machinery

Conditioning and temperature control are accomplished by recirculating reactants from storage tanks which are jacketed and/or contain tempering coils to maintain the process temperature required by the chemical system.

RIM parts (especially polyurethanes) are usually removed from the mold before the chemical reactions that develop the physical properties are complete. The part is placed on a support jig that holds it in its final shape until it is fully cured. In some cases this is done by simply setting the supported part aside for 12–24 h. More often the supported part is postcured in an oven for several hours at temperatures of about 180°F (82°C). Nylons are completely reacted in the mold and postcuring is not necessary.

RIM polyurethanes made with aromatic isocyanates (such as pure or polymeric MDI) have a tendency to darken as a result of the effect of UV light on the chemical ring structure of the MDI component. Soft white limestone or fine carbon black is often used as filler to mask the effect of this color change.

Polyurethanes manufactured with aliphatic isocyanates are light stable, and products are molded in a wide range of bright colors. Especially interesting is the development of equipment to add color concentrate, usually dispersed in a polyol, directly into the mixing head attached to a given mold. The basic urethane formula is adjusted to compensate for the additional reactive polyol. Using this technique with a multiclamp RIM line, it is possible to mold different colors using a single RIM machine. In some cases, aromatic RIM systems are molded in color, then painted the same color. This technique eliminates the need to touch up every dent or scratch which would otherwise show up tan or white.

There are several systems for painting RIM parts. Usually, RIM parts are simply primed and painted. Before painting, however, the part is cleaned to remove mold-release agents. The most common mold releases are metal stearates (or soaps) that can be removed from the part by a water wash and paraffin waxes that are usually removed by solvent vapor degreasing. Silicone mold releases are to be avoided as they are very difficult to remove from the part, and paint will not stick to the silicone surface film.

1.14.2 Polyurethanes

The most common chemicals used in the RIM process for poyurethanes are isocyanates containing two or more isocyanate (–N=C=O) groups and polyols, which contain two or more hydroxyl (–OH) groups. These reactive end-groups, so named because they occur at the ends of the chemical structure, react chemically to form a urethane linkage:

$$(-\underset{\text{H}}{\overset{|}{\text{N}}}\!\!-\!\overset{\text{O}}{\overset{\|}{\text{C}}}-\text{O}-)$$

The chemical system, must be adjusted so that the number of isocyanates and hydroxyls balance and that all reactive end-groups are used in the formation of urethane linkages.

The number of polymer structures that can be formed using the urethane reaction is quite large. There are ways to produce polyurethanes having different physical properties (see "Polyurethanes" in Chapter 1 of *Industrial Polymers, Specialty Polymers, and Their Applications*). If linear polyols are reacted with diisocyanates, a flexible polyurethane will be formed. If a low boiling liquid, such as Refrigerant-11 (R-11), is incorporated into the system, the heat of reaction will produce a cellular structure. The resulting product will be flexible polyurethane foam.

The physical properties of these materials can be varied by selecting polyols with shorter or longer polyol chains. The most common polyol or macroglycol chains are polyethers and polyesters (Chapter 1 of *Industrial Polymers, Specialty Polymers, and Their Applications*). The composition of these thermoplastic chains also plays a role in the physical properties of the end product. These chain segments in the block copolymer are often referred to as "soft" blocks, or segments, while the polyurethane segments formed by the reaction of diisocyante with glycol are referred to as "hard" blocks or segments.

In addition to changing the chain composition and length, the physical properties can be varied by blending up to approximately 10% of a long-chain triol (such as a triol adduct of ethylene oxide and propylene oxide with glycerol) into the basic resin system formulation. This produces branching in the "soft" segment of the block copolymer. Excessive triol modification may, however, diminish physical properties. Use of short-chain triols such as glycerol will produce cross-linking in the "hard" segments of the polymer chain. The formation of hard blocks and cross-linking in the hard block tend produce a stiffer, more rigid product. The hard blocks tend to be crystalline and reinforce the amorphous polymer, improving its strength.

1.14.3 Nylons

RIM nylons, like polyurethanes, form polymers very rapidly by the reaction of chemical end-groups. The linkages produced are as follows:

$$\text{Polyesteramide prepolymer} + \underbrace{(CH_2)_5\ CONH}_{\text{Caprolactam}} \longrightarrow \text{Nylon block copolymer}$$

Equipment used to manufacture RIM products must be extensively rebuilt to process RIM nylons. Because the viscosity of nylon RIM systems is low and the ingredients are quite reactive, leakage at the seals and the volumetric efficiency of the metering pumps (that is, the amount of material actually pumped divided by the volume displaced by the metering pumps) may cause problems, which require special attention. Hence, most manufacturers recommend having the machine designed specifically for nylon RIM systems.

The first commercial product made from nylon RIM was a front quarter panel (fender) for the Oldsmobile Omega Sport. Because of the excellent impact strength of nylon RIM, it has been used for bumper covers and automobile *fascia*. It also finds application in housings for business machines and electronics.

1.15 Structural Reaction Injection Molding

Structural reaction injection molding (SRIM) may be considered as a natural evolution of RIM. It is a very attractive composite manufacturing process for producing large, complex structural parts economically. The basic concepts of the SRIM process are shown in Figure 1.57. A preformed reinforcement is placed in a

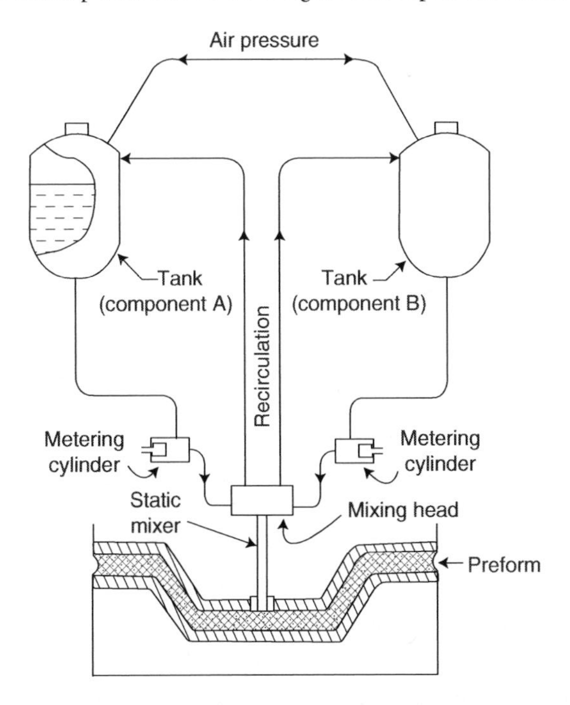

FIGURE 1.57 Schematic of structural reaction injection molding (SRIM).

closed mold, and a reactive resin mixture that is mixed by impingement under high pressure in a specially designed mix head (like that in RIM) flows at low pressure through a runner system to fill the mold cavity, impregnating the reinforcement material in the process. Once the mold cavity is filled, the resin quickly completes its reaction. A completed component can often be removed from the mold in as little as one minute.

SRIM is similar to RIM in its intensive resin mixing procedures and its reliance on fast resin reaction rates. It is also similar to resin transfer molding (RTM) (discussed later in this chapter) in employing preforms that are preplaced in the cavity of a compression mold to obtain optimum composite mechanical properties. The term *structural* is added to the term *RIM* to indicate the more highly reinforced nature of the composite components manufactured by SRIM.

The key to SRIM is the preform. It is a preshaped, three-dimensional precursor of the part to be molded and does not contain the resin matrix. It can consist of fibrous reinforcements, core materials, metallic inserts, or plastic inserts. The reinforcements, cores, or inserts can be anything available that meets the economic, structural, and durability requirements of the parts. This tremendous manufacturing freedom allows a variety of alternative preform constructions.

Most commercial SRIM applications have been in general industry or in the automotive industry. The reinforcement material most commonly used has been fiberglass, due to the low cost. Fiberglass has been used in the form of woven cloth, continuous strand mat, or chopped glass.

Space-shaping cores can be used in the SRIM process to fabricate thick, three-dimensional parts with low densities. Specific grades of urethane-based foams, having densities of 6–8 lb/ft^3 and dimensional stability at SRIM molding temperatures, are commonly used as molded core materials. Fiberglass reinforcements and inserts can be placed around these cores, resulting in SRIM parts, molded in one piece, that are very light-weight and structurally strong and stiff. Metallic inserts can be used in SRIM parts as local stiffeners, stressed attachment points, or weldable studs. The metallic material of choice is usually steel.

SRIM is a very labor-intensive process, and the consistency from preform to preform is usually poor. However, for very low manufacturing volumes this process can be cost-effective.

Most SRIM resins have several characteristics in common: their liquid reactants have room-temperature viscosity below 200 cps; their viscosity-cure curves are sigmoidal in shape, the typical mold-fill time being 10–90 sec; and their demold time is from 60 to 180 sec, varying with catalyst concentration. The low viscosity of SRIM resins and their relatively long fill times are crucial in allowing them to penetrate and flow through their reinforcing performs.

The design of the gating and runner configuration (if any) is usually kept proprietary by the molder. However, it appears that most SRIM parts are center-gated, with vents located along the periphery of the part. This configuration allows the displaced air in the mold cavity to be expelled uniformly.

1.15.1 Applications

The ability of SRIM to fabricate large, lightweight composite parts, consisting of all types of precisely located inserts and judiciously selected reinforcements, is an advantage that other manufacturing processes find difficult to match. Moreover, large SRIM parts can often be molded in 2–3 min, using clamping pressures as low as 100 psi. The capital requirements of SRIM are thus relatively low.

The first commercially produced SRIM part was the cover of the spare-tire well in several automobiles produced by General Motors. Since then, SRIM automotive structural parts have included foamed door panels, instrument panel inserts, sunshades, and rear window decks. Nonautomotive applications include satellite dishes and seat shells for the furniture market.

1.16 Resin Transfer Molding

RTM is similar to SRIM. In its common form, RTM is a closed-mold, low-pressure process in which dry, preshaped reinforcement material is placed in a closed mold and a polymer solution or resin is injected at a low pressure, filling the mold and thoroughly impregnating the reinforcement to form a composite part. The mold pressure in the RTM process is lower than in both SRIM and RIM/RRIM and the molding cycle time is much longer. The reinforcement and resin may take many forms, and the low pressure combined with the preoriented reinforcement package, affords a large range of component sizes, geometries, and performance options.

RTM is an excellent process choice for making prototype components. It allows representative prototypes to be molded at low cost, unlike processes such as compression molding and injection molding, which require tools and equipment approaching actual production level.

When prototyping with RTM, less reactive resins are generally used, allowing long fill times and easier control of the vents. Sizes can range from small components to very large, complex, three-dimensional structures. RTM provides two finished surfaces and controlled thickness, while other processes used for prototyping, such as hand lay-up and wet molding, give only a single finished surface.

1.17 Foaming Processes

Plastics can be foamed in a variety of ways. The foamed plastics, also referred to as cellular or expanded plastics, have several inherent features which combine to make them economically important. Thus, a foamed plastic is a good heat insulator by virtue of the low conductivity of the gas (usually air) contained in the system, has a higher ratio of flexural modulus to density than when unfoamed, has greater load-bearing capacity per unit weight, and has considerably greater energy-storing or energy-dissipating capacity than the unfoamed material. Foamed plastics are therefore used in the making of insulation, as core materials for load-bearing structures, as packaging materials used in product protection during shipping, and as cushioning materials for furniture, bedding, and upholstery.

Among those plastics which are commercially produced in cellular form are polyurethane, PVC, polystyrene, polyethylene, polypropylene, epoxy, phenol-formaldehyde, urea-formaldehyde, ABS, cellulose acetate, styrene-acrylonitrile, silicone, and ionomers. However, note that it is possible today to produce virtually every thermoplastic and thermoset material in cellular form. In general, the basic properties of the respective polymers are present in the cellular products except, of course, those changed by conversion to the cellular form.

Foamed plastics can be classified according to the nature of cells in them into closed-cell type and open-cell type. In a closed-cell foam each individual cell, more or less spherical in shape, is completely closed in by a wall of plastic, whereas in an open-cell foam individual cells are inter-connecting, as in a sponge. Closed-cell foams are usually produced in processes where some pressure is maintained during the cell formation stage. Free expansion during cell formation typically produces open-cell foams. Most foaming processes, however, produce both kinds.

A closed-cell foam makes a better buoy or life jacket because the cells do not fill with liquid. In cushioning applications, however, it is desirable to have compression to cause air to flow from cell to cell and thereby dissipate energy, so the open-cell type is more suitable. Foamed plastics can be produced in a wide range of densities—from 0.1 lb/ft.^{3} (0.0016 g/cm^3) to 60 lb/ft.^{3} (0.96 g/cm^3)—and can be made flexible, semirigid, or rigid.

A rigid foam is defined as one in which the polymer matrix exists in the crystalline state or, if amorphous, is below its T_g. Following from this, a flexible cellular polymer is a system in which the matrix polymer is above its T_g. According to this classification, most polyolefins, polystyrene, phenolic, polyycarbonate, polyphenylene oxide, and some polyurethane foams are rigid, whereas rubber foams, elastomeric polyurethanes, certain polyolefins, and plasticized PVC are flexible. Intermediate between these two extremes is a class of polymer foams known as semirigid. Their stress-strain behavior is, however, closer to that of flexible systems than to that exhibited by rigid cellular polymers.

The group of rigid cellular polymers can be further subdivided according to whether they are used (1) for non-load-bearing applications, such as thermal insulation; or as (2) load-bearing structural materials, which require high stiffness, strength and impact resistance.

The description of cellular foams as low, medium or high density is very common in practice. This is, however, not exact as the different density ranges which correspond to each of these items are not strictly defined. The following figures can, however, serve as a rough general guide:

	lb/ft.^{3}	Kg/m^3
Low density	0.1–3	2–50
Medium density	3–21	50–350
High density	21–60	350–960

(Note that the density of a polymer foam refers to its bulk density, defined by the ratio of total weight/total volume of the polymer and gaseous component. Obviously the gas phase contributes considerably to the volume of the end product, while the solid component contributes almost to the entire weight.)

Obtained forms of foamed plastics are blocks, sheets, slabs, boards, molded products, and extruded shapes. These plastics can also be sprayed onto substrates to form coatings, foamed in place between walls (i.e., poured into the empty space in liquid form and allowed to foam), or used as a core in mechanical structures. It has also become possible to process foamed plastics by conventional processing machines like extruders and injection-molding machines.

Polymer foams may be homogeneous with a uniform cellular morphology throughout or they may be structurally anisotropic. They may have an integral solid polymer skin or they may be multicomponent in which the polymer skin is of different composition to the polymeric cellular core. Schematic representations of the different physical forms of cellular polymers are given in Figure 1.58. Some

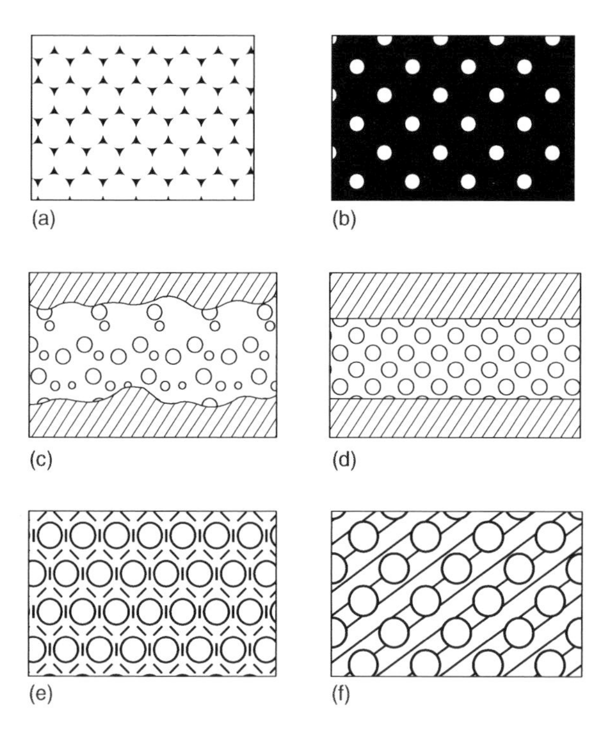

FIGURE 1.58 Schematic representations of section through different types of cellular polymer. (a) Low-density open-cell foam. (b) High-density closed-cell foam. (c) Single-component structural foam with cellular core and integral solid skin. (d) Multicomponent structural foam. (e) Fiber-reinforced closed-cell foam. (f) Syntactic foam.

special types of foams, namely, structural foams, reinforced foams, and syntactic foams are represented by Figure 1.58c to Figure 1.58f. These are described in a later section.

Foaming of plastics can be done in a variety of ways. Most of them typically involve creating gases to make foam during the foaming cycle. Once the polymer has been expanded or "blown," the cellular structure must be stabilized rapidly; otherwise it would collapse. Two stabilization methods are used. First, if the polymer is a thermoplastic, expansion is carried out above the softening or melting point, and the form is then immediately cooled to below this temperature. This is called physical stabilization. The second method—chemical stabilization—requires the polymer to be cross-linked immediately following the expansion step. Common foaming processes are the following:

1. Air is whipped into a dispersion or solution of the plastic, which is then hardened by heat or catalytic action or both.
2. A low-boiling liquid is incorporated in the plastic mix and volatilized by heat.
3. Carbon dioxide gas is produced within the plastic mass by chemical reaction.
4. A gas, such as nitrogen, is dissolved in the plastic melt under pressure and allowed to expand by reducing the pressure as the melt is extruded.
5. A gas, such as nitrogen, is generated within the plastic mass by thermal decomposition of a chemical blowing agent.
6. Microscopically small hollow beads of resin or even glass (e.g., microballoons) are embedded in a resin matrix.

Foams can be made with both thermoplastic and thermosetting plastics. The well known commercial thermoplastic foams are polystyrene, PVC, polyethylene, polypropylene, ABS copolymer, cellulose acetate. The thermosetting plastics which may be mentioned, among others, are phenol-formaldehyde, urea-formaldehyde, polyurethane, epoxy, and silicone. The methods of manufacture of some of these polymeric foams are given below.

1.17.1 Rigid Foam Blowing Agents

There are four types of polymers typically used for rigid foam production, namely, polystyrene, polyurethane, polyolefin, and phenolic. Within the polyolefin segment, rigid foams can be produced using polyethylene or polypropylene. Following the implementation of the Montreal Protocol, chlorofluorocarbons (CFC-11 and CFC-12) which had been the primary blowing agents for both flexible and rigid foams, were no longer available. Hydrochlorofluorocarbons (HCFCs) were one of the primary blowing agents that were then adopted, specifically HCFC-141b, HCFC-142b, and HCFC-22. Insulating foam products (with some exceptions) generally utilize HCFCs due to the superior insulation properties that they impart. The non-ozone depleting (i.e., not CFCs or HCFCs) blowing agents that are currently in use and will be substituted for HCFCs as the latter are phased out are: (i) hydrofluorocarbons (HFCs); (ii) hydrocarbons (e.g., pentanes, butanes); and (iii) carbon dioxide.

Non-insulating foam products typically utilize hydrocarbons, such as isobutane, pentane, isopentane, and hexane. The use of CO_2 (either water-based or liquid) is a major identified option to reduce the emission of non-HCFC blowing agents from polyurethane foam and extruded polystyrene boardstock applications. However, the thermal insulation properties of CO_2-blown foam are significantly compromised when compared to halocarbon-blown foam. Halocarbons (i.e., HCFCs, HFCs) are thus expected to be used in insulation foam manufacture for several years into the future. The primary HCFC replacements in these sectors are expected to be the liquid HFCs, which may see extensive use once HCFCs can no longer be used.

1.17.2 Polystyrene Foams

Polystyrene, widely used in injection and extrusion molding, is also extensively used in the manufacture of plastic foams for a variety of applications. Polystyrene produces light, rigid, closed-cell plastic foams having low thermal conductivity and excellent water resistance, meeting the requirements of low-temperature insulation and buoyancy applications. Two types of low-density polystyrene foams are available to the fabricator, molder, or user: (1) extruded polystyrene foam and (2) expandable polystyrene for molded foam.

1.17.2.1 Extruded Polystyrene Foam

This material is manufactured as billets and boards by extruding molten polystyrene containing a blowing agent (nitrogen gas or chemical blowing agent) under elevated temperature and pressure into the atmosphere where the mass expands and solidifies into a rigid foam. Many sizes of extruded foam are available, some as large as 10 in. × 24 in. × 9 ft. The billets and boards can be used directly or cut into different forms. One of the largest markets for extruded polystyrene in the form of boards is in low-temperature insulation (e.g., truck bodies, railroad cars, refrigerated pipelines, and low-temperature storage tanks for such things as liquefied natural gas). Another growing market for extruded polystyrene boards is residential insulation. Such boards are also used as the core material for structural sandwich panels, used prominently in the construction of recreational vehicles.

1.17.2.2 Expandable Polystyrene

Expandable polystyrene is produced in the form of free-flowing pellets or beads containing a blowing agent. Thus, pellets chopped from an ordinary melt extruder or beads produced by suspension polymerization are impregnated with a hydrocarbon such as pentane. Below 70°F (21°C) the vapor pressure of the pentane dissolved in the polymer is low enough to permit storage of the impregnated material (expandable polystyrene) in a closed container at ordinary temperature and pressure. Even so, manufacturers do not recommend storing for more than a few months.

The expandable polystyrene beads may be used in a tabular blow-extrusion process (Figure 1.59) to produce polystyrene foam sheet, which can subsequently be formed into containers, such as egg cartons and cold-drink cups, by thermoforming techniques.

Expandable polystyrene beads are often molded in two separate steps: (1) Preexpansion or prefoaming of the expandable beads by heat, and (2) further expansion and fusion of the preexpanded beads by heat in the enclosed space of a shaping mold.

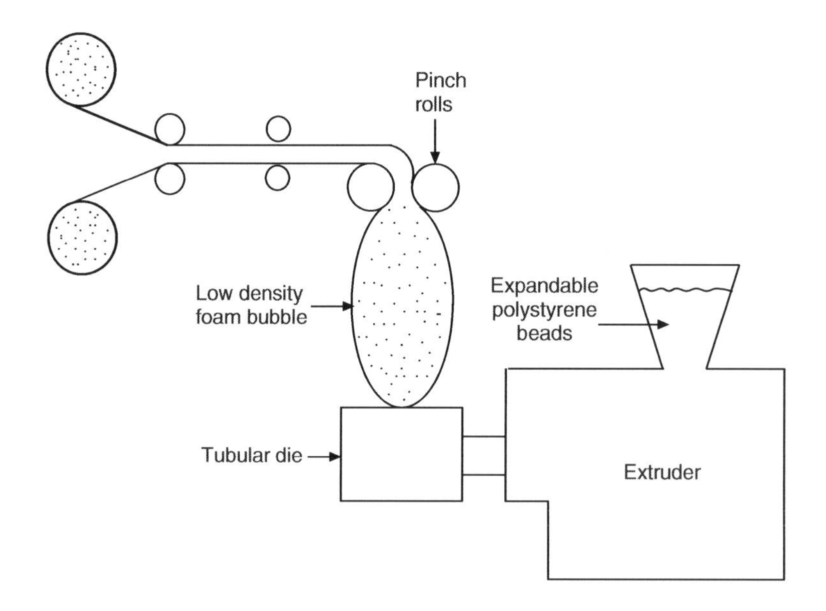

FIGURE 1.59 Tubular blow extrusion for production of low-density polystyrene foam sheet.

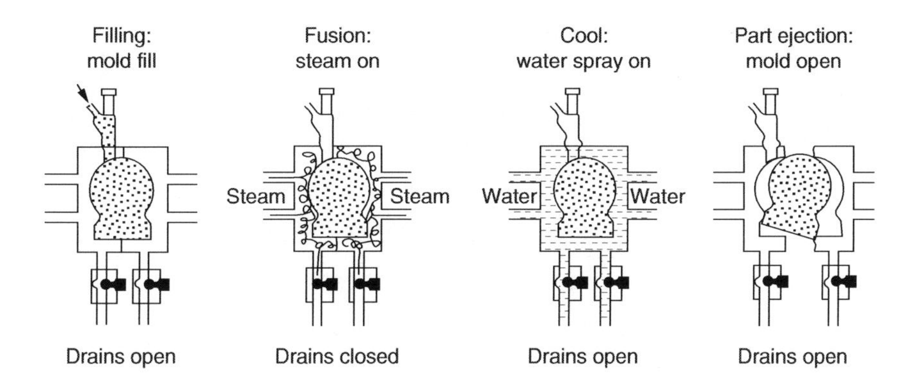

FIGURE 1.60 Molding of preexpanded (prefoamed) polystyrene beads. (Adapted from *PELASPAN Exapandable Polystyrene*, Form 171–414, Dow Chemical Co., 1966.)

Steam heat is used for preexpansion in an agitated drum with a residence time of a few minutes. As the beads expand, a rotating agitator prevents them from fusing together, and the preexpanded beads, being lighter, are forced to the top of the drum and out the discharge chute. They are then collected in storage bins for aging prior to molding. The usual lower limit of bulk density for bead preexpansion is 1.0 lb/ft.^{3} (0.016 g/cm^3), compared to the virgin bead bulk density of about 35 lb/ft.^{3} (0.56 g/cm^3).

Molding of preexpanded (prefoamed) beads requires exposing them to heat in a confined space. In a typical operation (Figure 1.60) prefoamed beads are loaded into the mold cavity, the mold is closed, and steam is injected into the mold jacket. The prefoamed beads expand further and fuse together as the temperature exceeds T_g. The mold is cooled by water spray before removing the molded article. Packages shaped to fit their contents (small sailboats, toys, drinking cups, etc.) are made in this way. Special machines have been designed to produce thin-walled polystyrene foam cups. Very small beads at a prefoamed density of approximately 4–5 lb/ft.^{3} (0.06–0.08 g/cm^3) are used, which allow easy flow into the molding cavity and produce a cup having the proper stiffness for handling.

1.17.2.3 Structural Foams

Structural foam is the term usually used for foam produced in an injection molding press and made of almost many thermoplastic resin. Structural foam is always produced with a hard integral skin on the outer surfaces and a cellular core in the interior, and is used almost exclusively for production of molded parts. The process is thus ideally suited for fabrication of parts such as business machine housings (commonly for ABS), and similar parts or components in which lightweight and stiffness are required.

The structural foam injection molding process (Figure 1.61), by which a product with a cellular core and a solid skin can be molded in a single operation, gets its name from the application of its product rather than the mechanism of the process itself. In a manner directly opposite to the vented extruder (Figure 1.21), a blowing agent, often nitrogen, is injected into the melt in the extruder. The polymer melt, injected with gas, is then forced into the accumulator where it is maintained at a pressure and temperature high enough to prevent foaming (Figure 1.61a). When a sufficient charge has accumulated it is transferred into the mold (Figure 1.61b). The melt foams and fills the mold at a relatively low pressure (1.3–2.6 MPa) compared to the much higher pressure in the accumulator. The lower operating pressures of the molds make the molds less expensive than those used for conventional injection molding. However, the cycle times are longer because the foam being a good insulator, takes longer time to cool.

Structural foams can also be made using a chemical blowing agent (discussed later) rather than an inert gas. In that case, a change in pressure or temperature on entering the mold triggers gas formation. Today structural foam injection molding is a very fast-growing polymer processing technique that can be used to modify the properties of thermoplastics to suit specific applications.

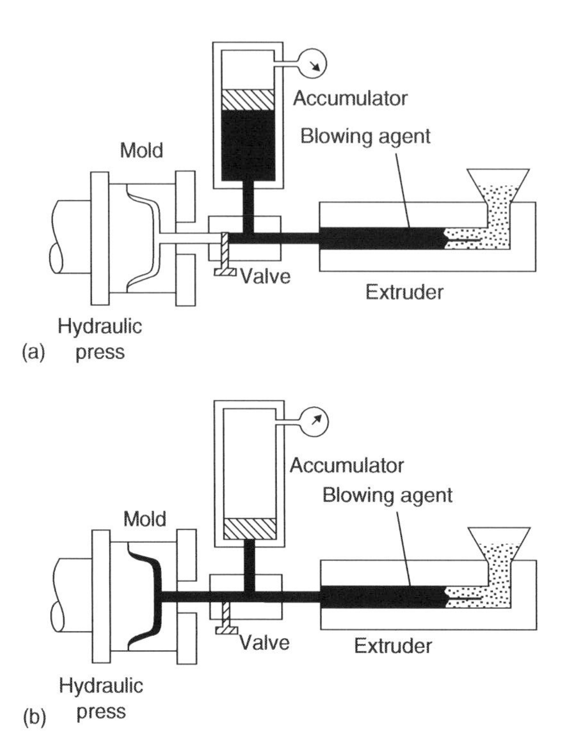

FIGURE 1.61 Structural foam process. (a) Filling the accumulator. The blowing agent (usually nitrogen) is injected into the melt in the extruder before it is passed into the accumulator. (b) Filling the mold. The accumulator ram injects the melt into the mold where the reduced pressure allows the gas to foam the resin.

1.17.3 Polyolefin Foams

Polyolefin foams can be produced with closely controlled density and cell structure. Generally the mechanical properties of polyolefins lies between those of a rigid and a flexible foam. Polyolefin foams have a very good chemical and abrasion resistance as well as good thermal insulation properties. Cross-linking improves foam stability and polymer properties.

A variety of foams can be produced from various types of polyethylenes and cross-linked systems having a very wide range of physical properties, and foams can be tailor-made to a specific application. Polypropylene has a higher thermostability than polyethylene. The production volume of polyolefin foams is not as high as that of polystyrene, polyurethane, or PVC foams. This is due to the higher cost of production and some technical difficulties in the production of polyolefin foams. The structural foam injection molding process, described previously for polystyrene, is also used for polyethylene and polypropylene structural foams (see Figure 1.61).

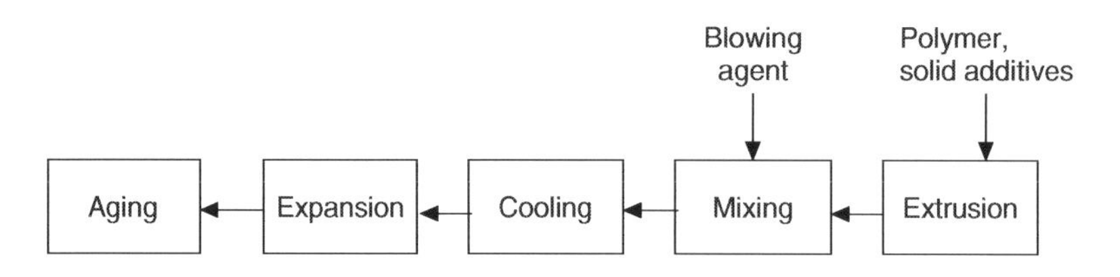

FIGURE 1.62 Block diagram of polyolefin foam manufacture by extrusion process.

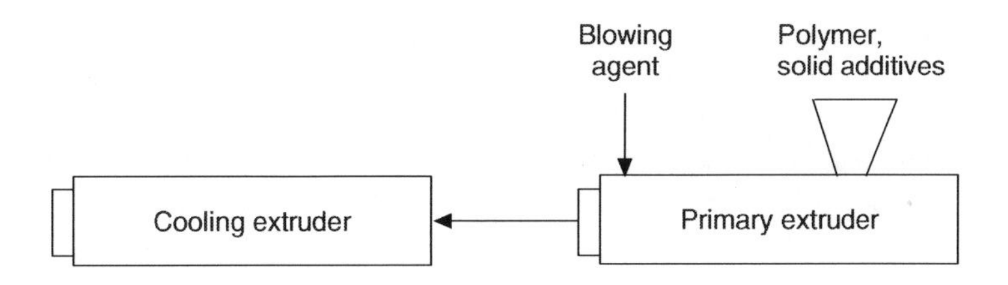

FIGURE 1.63 Schematic diagram of tandem extruder.

Commercial extrusion processes for polyolefin foam products are derived from the original Dow process which basically involves five steps, namely, extrusion, mixing, cooling, expansion, and aging (see Figure 1.62). These steps of the extrusion process may be performed on equipment of several different configurations such as single-screw extruders, twin-screw extruders, and tandem-extruder lines. Single-screw extruders must be equipped with a multistage long screw of high length-to-diameter ratio capable of performing all the aforesaid extrusion steps. Twin-screw extruders, on the other hand, have low shear rate and high mixing ability both of which are desirable in foam extrusion.

Except where the extrusion rate is low such as for products having a small cross-section, most polyolefin foam products are made with tandem-type extruders, as shown in Figure 1.63. The primary extruder consisting of a two-stage screw melts the resin and then mixes the melt with the solid additives and liquid blowing agent, whereas the second extruder, usually larger than the primary one and designed to provide maximum cooling efficiency, cools the molten polymer mixture to the optimum foaming temperature. In some equipment, however, the second extruder is designed to perform both as mixer for the blowing agent and as cooler.

An alternative to large tandem extruders is the accumulating extrusion system which provides a high instantaneous extrusion rate. It is commonly employed for producing large plank products. In the simple system, shown in Figure 1.64, the foamable melt is fed into an accumulator by a single screw extruder and pushed out by a ram through a die orifice. The process is discontinuous, resulting in a loss of yield, but it has the advantage of low capital requirement.

The shape of a polyolefin foam product is determined largely by the shape of the die. Thus a circular die is used for a rod, an annular die for a tube or sheet product, and a slit die for a plank product. To make

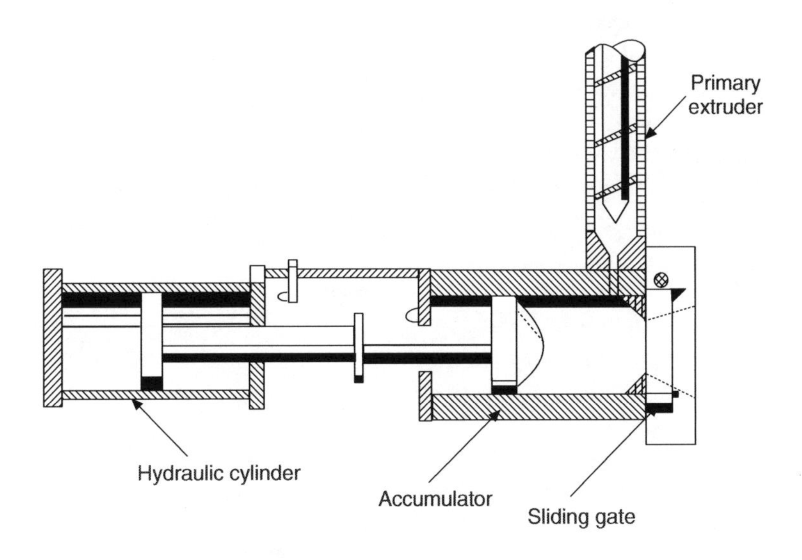

FIGURE 1.64 Schematic of accumulation extrusion system.

polyolefin foam sheet, the extruded tubular foam, expanding at the annular die, is guided over a sizing mandrel, slit, laid flat, and wound into a roll [33]. Cool air is blown in at the nose of the mandrel to reduce the friction between the hot expanding foam and the mandrel.

The extruded polyolefin foam must be dimensionally stabilized by aging, since the foam deforms according to the internal cell pressure, which changes with time as air and gaseous blowing agent diffuse into and out of the foam at different rates. If the rates are equal, the cell pressure, and hence the foam dimensions, will remain constant, as is found for the LDPE/CDC-114 system. Most blowing agents, however, permeate through polyolefins faster than air and, as a result, the foams shrink. The aging time required for the shrunken foam to recover and stabilize depends on the properties of the polymer and the physical attributes of the foam, such as the open-cell content, foam density, and foam thickness. The aging time may range from less than a week for a thin sheet to several weeks for a thick plank.

The production line of a typical process to manufacture thin ultra-low-density (ULD) polypropylene foam sheet, consists of tandem extruders, an accumulating vessel, and an annular die. The secondary extruder is designed to mix a large amount of blowing agent into the polymer and then to cool the mixture. The blowing agent consists of a large proportion (90%) of a highly soluble blowing agent such as CFC-11 which provides the heat sink necessary for foam stabilization and a small proportion (10%) of a low-permeability blowing agent such as CFC-12 or CFC-114 which serves as an inflatant. The accumulating extrusion system allows the high extrusion rate required for the production of ULD foam sheet.

There are several processes for the production of moldable polyolefin beads. In the BASF process, LDPE foam strands are extruded out of a multi-hole die and granulated to beads by a die-face cutter. Inexpensive butane is used as the blowing agent and the foam beads are then cross-linked by electron-beam. As the beads have atmospheric cell pressure, a special technique is required to develop the necessary cell pressure for molding [34].

In the Kanegafuchi process, most widely used to manufacture LDPE foam beads, dicumyl peroxide is impregnated into finely pelletized LDPE beads suspended in water in an autoclave with the help of a dispersant such as basic calcium tertiary phosphate and sodium dodecylbenzene sulfonate. The beads are then heated to cross-link. The cross-linked beads are impregnated with a blowing agent (e.g., CFC-12), cooled, discharged from the autoclave and immediately expanded with steam to make foam beads. For molding, the foam beads are charged into a mold and heated with superheated steam (>140 kPa) to expand and weld.

The majority of cross-linked polyolefin foam sheet products are made by one of the four Japanese processes: Sekisui, Toray, Furukawa, and Hitachi, the first two of which use the radiation method and the latter two a chemical method for cross-linking.

The flow diagram of the radiation cross-linked polyolefin foam sheet process is shown in Figure 1.65. The key steps of the process include a uniform mixing of polymer and blowing agent (powder), manufacturing void-free sheet of uniform thickness, cross-linking the sheet to the desired degree by irradiation with a high-energy ray, and then softening and expanding the sheet in a foaming chamber

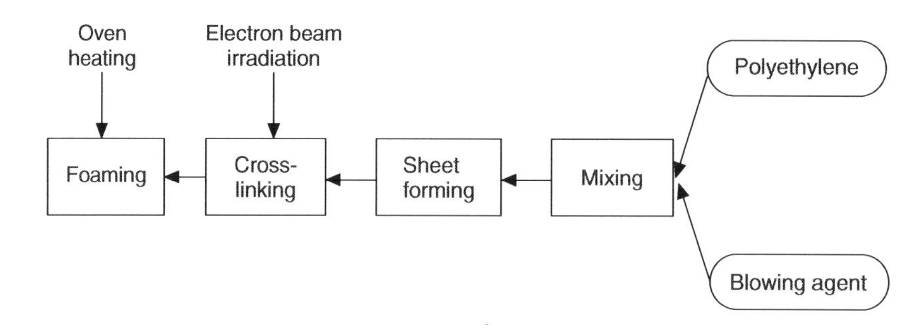

FIGURE 1.65 Flow diagram of radiation cross-linked polyolefin foam sheet forming process.

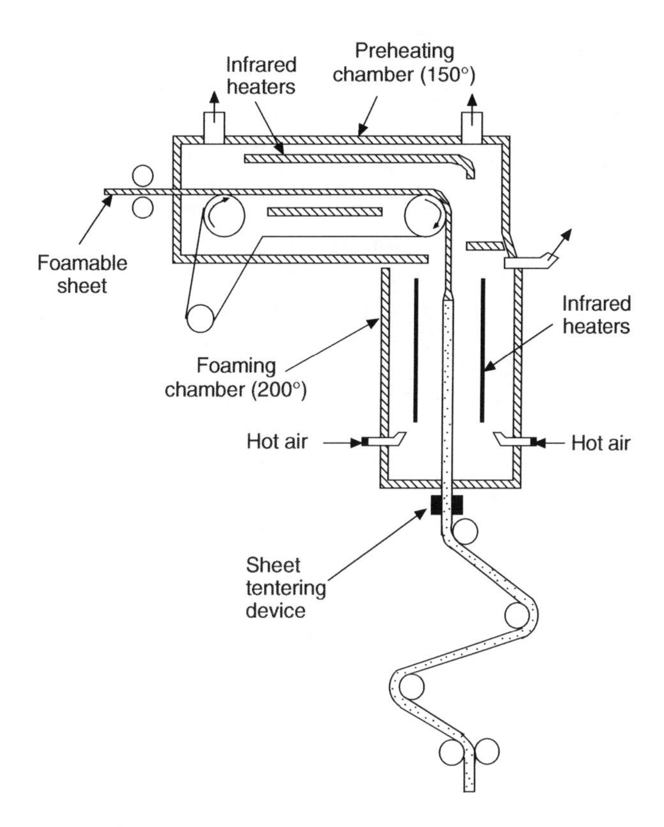

FIGURE 1.66 Schematic of Sekisui vertical foaming oven for cross-linked polyolefin foam sheet.

(oven) using a suitable support mechanism. The Sekisui process employs a vertical air oven like the one shown in Figure 1.66 for expanding the foamable sheet. The oven consists of a horizontal preheating chamber and a vertical foaming chamber. The rapidly expanding sheet supports itself by gravity in the vertical direction, while a specially designed tentering device keeps the sheet spread out. In the Toray process, the foamable sheet is expanded while afloat on the surface of molten salts. The process is suitable for producing cross-linked PP foam sheet as well as PE foam sheet.

The flow diagram of the chemically cross-linked polyolefin foam sheet process is shown in Figure 1.67. Unlike in the radiation cross-linking process, a peroxide cross-linking agent is incorporated in the polymer along with the blowing agent. Therefore, a tighter temperature control must be maintained in the sheet manufacturing steps to prevent premature cross-linking by the peroxide. In the oven, on the other hand, the cross-linking of the polyolefin sheet must be thermally effected without causing the

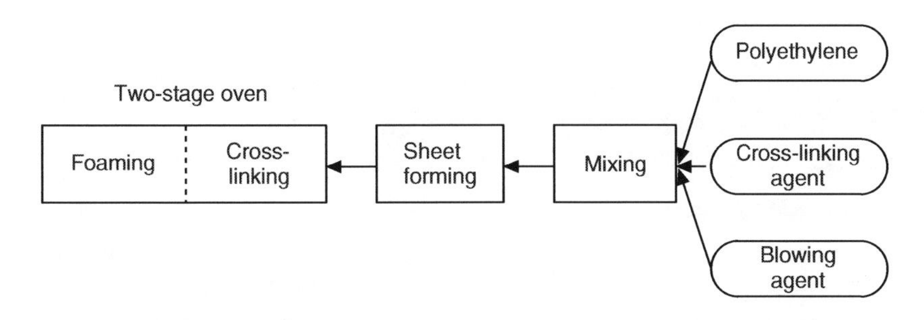

FIGURE 1.67 Flow diagram of chemically cross-linked polyolefin foam sheet forming process.

blowing agent to decompose. Consequently, both the oven design and the selection of raw materials are more difficult in the chemical cross-linking process. Both the Furukawa and Hitachi processes employ horizontal air ovens consisting of at least two sections, the preheating section and the foaming/forming section, and having one or more non-stick conveyors to support the sheet during heating and expansion [33].

Polyolefin foams have many and varied applications due to their unique properties which include buoyancy, resiliency, energy absorption, low thermal conductivity, resistance to chemicals, thermoformability, and ease of fabrication. The major application areas of polyolefin foams are cushion packaging (pads and saddles, encapsulation, case inserts, etc.), construction (expansion joint filler, closure strips, floor underlayment etc.), automotive (headliner, door trim, instrument panel, trunk liner, air conditioner liner, etc.), insulation (insulation of pipe, storage tanks), and sports and leisure (life vests, surfboards, swim aids, ski belts, gym mats, etc.). Thin polyolefin foam sheet products are used primarily as wrapping materials to protect the surfaces of articles from minor dents and abrasion during handling and shipping.

1.17.4 Polyurethane Foams

Polyurethane foams, also known as urethane foams or U-foams, are prepared by reacting hydroxyl-terminated compounds called polyols with an isocyanate (see Figure 1.29 of *Plastics Fundamentals, Properties, and Testing*). Isocyanates in use today include toluene diisocyanate, known as TDI, crude methylenebis(4-phenyl-isocyanate), known as MDI, and different types of blends, such as TDI/crude MDI. Polyols, the other major ingredient of the urethane foam, are active hydrogen-containing compounds, usually polyester diols and polyether diols.

It is possible to prepare many different types of foams by simply changing the molecular weight of the polyol, since it is the molecular backbone formed by the reaction between isocyanate and polyol that supplies the reactive sites for cross-linking (Figure 1.29 of *Plastics Fundamentals, Properties, and Testing*), which in turn largely determines whether a given foam will be flexible, semirigid, or rigid. In general, high-molecular-weight polyols with low functionality produce a structure with a low amount of cross-linking and, hence, a flexible foam. On the other hand, low-molecular-weight polyols of high functionality produce a structure with a high degree of cross-linking and, consequently, a rigid foam. Of course, the formulation can be varied to produce any degree of flexibility or rigidity within these two extremes.

The reactions by which urethane foam are produced can be carried out in a single stage (one-shot process) or in a sequence of several stages (prepolymer process and quasi-prepolymer process.) These variations led to 27 basic types of products or processes, all of which have been used commercially.

In the one-shot process, all of the ingredients—isocyanate, polyol, blowing agent, catalyst, additives, etc.—are mixed simultaneously, and the mixture is allowed to foam. In the prepolymer method (Figure 1.29 of *Plastics Fundamentals, Properties, and Testing*), a portion of the polyol is reacted with an excess of isocyanate to yield a prepolymer having isocyanate end groups. The prepolymer is then mixed with additional polyol, catalyst, and other additives to cause foaming. The quasi-prepolymer process is intermediate between the prepolymer and one-shot processes.

1.17.4.1 Flexible Polyurethane Foams

The major interest in flexible polyurethane foams is for cushions and other upholstery materials. Principal market outlets include furniture cushioning, carpet underlay, bedding, automotive seating, crash pads for automobiles, and packaging. The density range of flexible foams is usually 1–6 lb/ft.^{3} (0.016–0.096 g/cm^3). The foam is made in continuous loaves several feet in width and height and then sliced into slabs of desired thickness.

1.17.4.1.1 One-Shot Process

The bulk of the flexible polyurethane foam is now being manufactured by the one-shot process using polyether-type polyols because they generally produce foams of better cushioning characteristics. The main components of a one-shot formulation are polyol, isocyanate, catalyst, surfactant, and blowing agent.

Today the bulk of the polyether polyols used for flexible foams are propylene oxide polymers. The polymers prepared by polymerizing the oxide in the presence of propylene glycol as an initiator and a caustic catalyst are diols having the general structure

$$CH_3-\underset{\displaystyle OH}{\underset{|}{CH}}-CH_2-O\!\left(CH_2-\underset{\displaystyle CH_3}{\underset{|}{CH}}-O\right)_{\!n}CH_2-\underset{\displaystyle OH}{\underset{|}{CH}}-CH_3$$

The polyethers made by polymerizing propylene oxide using trimethylol propane, 1,2,6-hexanetriol, or glycerol as initiator are polymeric triols. For example, glycerol gives

$$HO\!\left(C_3H_6O\right)_{\!n}CH_2-CH(OH)-CH_2\!\left(C_3H_6O\right)_{\!n}OH$$

The higher hydroxyl content of these polyethers leads to foams of better loadbearing characteristics. Molecular weight in the range 3,000–3,500 is found to give the best combination of properties.

The second largest component in the foam formulation is the isocyanate. The most suitable and most commonly used isocyanate is 80:20 TDI—i.e., 80:20 mixture of tolylene-2,4-diisocyanate and tolylene-2,6-diisocyanate.

One-shot processes require sufficiently powerful catalysts to catalyze both the gas evolution and chain extension reaction (Figure 1.29 of *Plastics Fundamentals, Properties, and Testing*). Use of varying combinations of an organometallic tin catalyst (such as dibutyltin dilaurate and stannous octoate) with a tertiary amine (such as alkyl morpholines and triethylamine), makes it possible to obtain highly active systems in which foaming and cross-linking reactions could be properly balanced.

The surface active agent is an essential ingredient in formulations. It facilitates the dispersion of water (see below) in the hydrophobic resin by decreasing the surface tension of the system. In addition, it also aids nucleation, stabilizes the foam, and regulates the cell size and uniformity. A wide range of surfactants, both ionic and nonionic, have been used at various times. Commonly used among them are the water-soluble polyether siloxanes.

Water is an important additive in urethane foam formulation. The water reacts with isocyanate to produce carbon dioxide and urea bridges (Figure 1.29 of *Plastics Fundamentals, Properties, and Testing*). An additional amount of isocyanate corresponding to the water present must therefore be incorporated in the foaming mix. The more water that is present, the more gas that is evolved and the greater number of active urea points for cross-linking. This results in foams of lower density but higher degree of cross-linking, which reduces flexibility. So when soft foams are required, a volatile liquid such as trichloromonofluoromethane (bp 23.8°C) may be incorporated as a blowing agent. This liquid will volatilize during the exothermic urethane reaction and will increase the total gas in the foaming system, thereby

TABLE 1.2 Urethane One-Shot Foam Formulation

Ingredient	Parts by Weight
Poly(propylene oxide), mol. wt. 2000 and 2OH/molecule	35.5
Poly(propylene oxide) initiated with trifunctional alcohol, mol. wt. 3000 and 3 OH/molecule	35.5
Toluene diisocyanate (80:20 TDI)	26.0
Dibutyltin dilaurate	0.3
Triethylamine	0.05
Water	1.85
Surfactant (silicone)	0.60
Trichloromonofluoromethane (CCl_3F)	12.0
Final density of foam = 1.4 lb/ft.^{3} or 0.022 g/cm^3 (2.0 lb/ft.^{3} or 0.032 g/cm^3 if CCl_3F is omitted)	

Source: One-Step Urethane Foams, 1959. Bull, F40487, Union Carbide Corp.

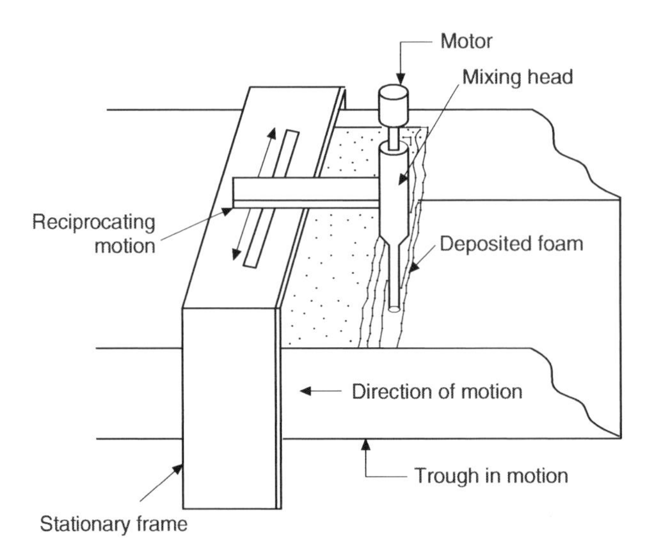

FIGURE 1.68 Schematic of a Hanecke-type machine for production of polyurethane foam in block form by one-shot process.

decreasing the density, but it will not increase the degree of cross-linking. However, where it is desired to increase the cross-link density independently of the isocyanate-water reaction, polyvalent alcohols, such as glycerol and pentaerythritol, and various amines may be added as additional cross-linking agents. A typical formulation of one-shot urethane foam system is shown in Table 1.2

Most foam is produced in block form from Henecke-type machines (Figure 1.68) or some modification of them. In this process [35], several streams of the ingredients are fed to a mixing head which oscillates in a horizontal plane. In a typical process, four streams may be fed to the mixing head: e.g., polyol and fluorocarbon (if any); isocyanate; water, amine, and silicone; and tin catalyst. The reaction is carried out with slightly warmed components. Foaming is generally complete within a minute of the mixture emerging from the mixing head. The emergent reacting mixture runs into a trough, which is moving backward at right angles to the direction of traverse of the reciprocating mixing head. In this way the whole trough is covered with the foaming mass.

Other developments of one-shot flexible foam systems include direct molding, where the mixture is fed into a mold cavity (with or without inserts such as springs, frames, etc.) and cured by heat. In a typical application, molds would be filled and closed, then heated rapidly to 300°F–400°F (149°C–204°C) to develop maximum properties.

A good deal of flexible urethane foam is now being made by the cold-cure technique. This involves more reactive polyols and isocyanates in special foaming formulations which would cure in a reasonable time to their maximum physical properties without the need for additional heat over and above that supplied by the exothermic reaction of the foaming process.

Cold-cure foaming is used in the production of what is known as high-resilient foams having high sag factor (i.e., ratio of the load needed to compress foam by 65% to the load needed to compress foam by 25%), which is most important to cushioning characteristics. True cold-cure foams will produce a sag factor of 3–3.2, compared to 2–2.6 for hot-cured foams.

1.17.4.1.2 Prepolymer Process

In the prepolymer process the polyol is reacted with an excess of isocyanate to give an isocyanate-terminated prepolymer which is reasonably stable and has less handling hazards than free isocyanate. If water, catalysts, and other ingredients are added to the product, a foam will result. For better load-bearing and cushioning properties, a low-molecular-weight triol, such as glycerol and trimethylolpropane, is added to the polyol before it reacts with the isocyanate. The triol provides a site for chain branching.

Although the two-step prepolymer process is less important than the one-shot process, it has the advantage of low exotherms, greater flexibility in design of compounds, and reduced handling hazards.

1.17.4.1.3 Quasi-Prepolymer Process

In the quasi-prepolymer process a prepolymer of low molecular weight and hence low viscosity is formed by reacting a polyol with a large excess of isocyanate. This prepolymer, which has a large number of free isocyanate groups, is then reacted at the time of foaming with additional hydroxy compound, water, and catalyst to produce the foam. The additional hydroxy compound, which may be a polyol or a simple molecule such as glycerol or ethylene glycol, also functions as a viscosity depressant. The system thus has the advantage of having low-viscosity components, compared to the prepolymer process, but there are problems with high exotherms and a high free-isocyanate content.

Quasi-prepolymer systems based on polyester polyols and polyether polyols are becoming important in shoe soling, the former being most wear resistant and the latter the easiest to process.

1.17.4.2 Rigid and Semirigid Foams

The flexible foams discussed in previous sections have polymer structures with low degrees of cross-linking. Semirigid and rigid forms of urethane are products having higher degree of cross-linking. Thus, if polyols of higher functionality—i.e., more hydroxyl groups per molecule—are used, less flexible products may be obtained, and in the case of polyol with a sufficiently high functionality, rigid foams will result.

The normal density range for rigid and semirigid foams is about 1–3 $lb/ft.^3$ (0.016–0.048 g/cm^3). Some packaging applications, however, use densities down to 0.5 $lb/ft.^3$ (0.008 g/cm^3); for furniture applications densities can go as high as 20–60 $lb/ft.^3$ (0.32–0.96 g/cm^3), thus approaching solids. At densities of from 2 $lb/ft.^3$ (0.032 g/cm^3) to 12 $lb/ft.^3$ (0.19 g/cm^3) or more, these foams combine the best of structural and insulating properties.

Semirigid (or semiflexible) foams are characterized by low resilience and high energy-absorbing characteristics. They have thus found prime outlet in the automotive industry for applications like safety padding, arm rests, horn buttons, etc. These foams are cold cured and involve special polymeric isocyanates. They are usually applied behind vinyl or ABS skins. In cold curing, the liquid ingredients are simply poured into a mold in which vinyl or ABS skins and metal inserts for attachments have been laid. The liquid foams and fills the cavity, bonding to the skin and inserts. Formulations and processing techniques are also available to produce self-skinning semirigid foam in which the foam comes out of the mold with a continuous skin of the same material.

Rigid urethane foams have outstanding thermal insulation properties and have proved to be far superior to any other polymeric foam in this respect. Besides, these rigid foams have excellent compressive strength, outstanding buoyancy (flotation) characteristics, good dimensional stability, low water absorption, and the ability to accept nails or screws like wood. Because of these characteristics, rigid foams have found ready acceptance for such applications as insulation, refrigeration, packaging, construction, marine uses, and transportation.

For such diverse applications several processes are now available to produce rigid urethane foam. These include foam-in-place (or pour-in-place), spraying, molding, slab, and laminates (i.e., foam cores with integral skins produced as a single unit). One-shot techniques can be used without difficulty, although in most systems the reaction is slower than with the flexible foam, and conditions of manufacture are less critical. Prepolymer and quasi-prepolymer systems were also developed in the United States for rigid and semirigid foams, largely to reduce the hazards involved in handling TDI where there are severe ventilation problems.

In the foam-in-place process a liquid urethane chemical mixture containing a fluorocarbon blowing agent is simply poured into a cavity or metered in by machine. The liquid flows to the bottom of the cavity and foams up, filling all cracks and corners and forming a strong seamless core with good adhesion to the inside of the walls that form the cavity. The cavity, of course, can be any space, from the space between two walls of a refrigerator to that between the top and bottom hull of a boat. However, if the cavity is the interior of a closed mold, the process is known as molding.

Rigid urethane foam can be applied by spraying with a two-component spray gun and a urethane system in which all reactants are incorporated either in the polyol or in the isocyanate. The spraying process can be used for applying rigid foam to the inside of building panels, for insulating cold-storage rooms, for insulating railroad cars, etc.

Rigid urethane foam is made in the form of slab stock by the one-shot technique. As in the Henecke process (Figure 1.61), the reactants are metered separately into a mixing head where they are mixed and deposited onto a conveyor. The mixing head oscillates in a horizontal plane to insure an even deposition. Since the foaming urethane can structurally bond itself to most substrates, it is possible (by metering the liquid urethane mixture directly onto the surface skin) to produce board stock with integral skins already attached to the surface of the foam. Sandwich-construction building panels are made by this technique.

1.17.5 Foamed Rubber

Although foamed rubber and foamed urethanes have many similar properties, the processes by which they are made differ radically. In a simple process a solution of soap is added to natural rubber (NR) latex so that a froth will result on beating. Antioxidants, cross-linking agents, and a foam stabilizer are added as aqueous dispersions.

Foaming is done by combined agitation and aeration with automatic mixing and foaming machines. The stabilizer is usually sodium silicofluoride (Na_2SiF_6). The salt hydrolyzes, yielding a silica gel which increases the viscosity of the aqueous phase and prevents the foam from collapsing. A typical cross-linking agent is a combination of sulfur and the zinc salt of mercaptobenzothiazole (accelerator). Cross-linking (curing or vulcanization) with this agent takes place in 30 min at 100°C.

When making a large article such as a mattress, a metal mold may be filled with the foamed latex and heated by steam at atmospheric pressure. After removing the foamed rubber article from the mold, it may be dewatered by compressing it between rolls or by centrifuging and by drying with hot air in a tunnel dryer. In foamed rubber formulation a part of the NR latex can be replaced by a synthetic rubber latex. One such combination is shown in Table 1.3.

1.17.6 Epoxy Resins

Any epoxy resin can be made foamable by adding to the formulation some agent that is capable of generating a gas at the curing temperature prior to gelation. Such foaming agents may be low-boiling liquids which vaporize on heating (e.g., CFCs such as Freons) or blowing agents that liberate a gas when heated above 70°C, such as 2,2′-azobis(isobutyronitrile) or sulfonylhydrazide, which decompose

TABLE 1.3 Foamed Rubber Formulation

Ingredient	Parts by Weight
Styrene-butadiene latex (65% solids)	123
Natural rubber latex (60% solids)	33
Potassium oleate	0.75
Sulfur	2.25
Accelerators	
Zinc diethyldithiocarbamate	0.75
Zinc salt of mercaptobenzothiazole	1.0
Trimene base (reaction product of ethyl chloride, formaldehyde, and ammonia)	0.8
Antioxidant (phenolic)	0.75
Zinc oxide	3.0
Na_2SiF_6	2.5

Source: Stern, H. J. 1967. *Rubber: Natural and Synthetic*, Palmerton, New York.

evolving nitrogen. A foaming gas can also be generated in situ by adding a blowing agent that reacts with amine (curing agent) to form the gas. A typical such system consists of an epoxy resin, a primary amine (hardener), and a hydrogen-active siloxane (blowing agent). The siloxane reacts with the amine, evolving hydrogen as a foaming gas.

Instead of using amines it is also possible to use other hydrogen-active hardeners such as phenols and carboxylic acids. The reaction of gas evolution occurs immediately after mixing resin and hardeners and before the mixture begins to cure in the mold. This is essential for the formation of closed-cell foam structure during the curing, which takes place under a definite expansion pressure against the mold wall, leading to formation of dense casting. In the production of expanded laminates of sandwich configuration, this yields very tough and impact-resistant structures.

Expanded materials with excellent high-temperature properties are obtained when cresol novolacs are used as hardeners. A typical formulation is based on mixtures of bisphenol A resins and epoxy novolac resins which are cured by cresol novolacs and accelerated by suitable nitrogen-containing agents.

Applications of epoxy foams can be categorized in three areas, namely, (1) unreinforced materials, (2) glass-fiber-reinforced materials, and (3) sandwich constructions [33]. Because of their light weight and absence of shrinkage, foamed epoxies are used in the production of large-scale patterns. Having excellent dielectric properties, epoxy foams find applications in electronics such as for casting and sealing electronic components like small transformers and capacitors, and in insulating cables.

The light weight properties of foamed epoxies are utilized in fiber-reinforced materials for which glass fiber mats and unidirectional rovings are most suitable, with the majority of applications involving sandwich constructions. A few practical examples are foamed epoxy windsurfing board, epoxy rotor blades for wind energy generators, and automotive spoilers.

1.17.7 Urea-Formaldehyde Foams

Urea-formaldehyde (UF) foams are basically two-component systems as the production of the foams requires mainly a UF resin and a foam stabilizing agent. The UF resin (see Chapter 1 of *Industrial Polymers, Specialty Polymers, and Their Applications*) is produced by the condensation of urea and formaldehyde in the range of mole ratios of 2: 3–1: 2 in the presence of alkaline catalysts (pH $\sim$8), which yield only short-chain oligomers, and under weakly acidic conditions (pH 4–6), which result in a higher molecular weight mixture of oligomers (solubilized by attached methylol groups). Though numerous substances have been proposed as foam-stabilizing agents for the commercial foam system, aqueous solutions of the sodium salts of dodecylbenzene sulfonic acid and dibutylnaphthalene sulfonic acid have proven to be of value. Aqueous solutions of strongly dissociating organic and inorganic acids with a pH range of 1–1.5 are used as hardening agents. However, phosphoric acid is preferred because of its negligible corrosive action. The hardening agent is preferentially added to the foam stabilizing agent and the concentration of the hardener is chosen so that the final foam gels within 30–90 sec at 20–25°C after mixing the components.

In a stationary process for making UF foams, a mixture of water and foam-stabilizing agent is introduced into a vessel equipped with tubular vanes. Foaming takes place upon feeding air and a urea-formaldehyde resin solution (see "Urea-Formaldehyde Resin" in Chapter 2 of *Industrial Polymers, Specialty Polymers, and Their Applications*). The generated foam is guided by the action of the tubular vanes to the outlet channel of the vessel from where the foam exits as a rectangular slab and is transported on a conveyor belt until the foam structure hardens sufficiently. Blocks are cut from the foam slab, dried at about 40°C for about 2 h, and then pressed mechanically into sheets of the required dimensions [33].

For on-site production of foam, the raw materials are transported by pumps into a foaming machine. A dispersion of foam stabilizing agent is formed in water in the machine and the resin is introduced into the mixing chamber through jets. The finished foam emerges from the plastic pipe and can be used immediately at the site. One thousand liters (1 m^3) of foam can be produced on-site from 20 liters of resin and 18 liters of foam solution.

The UF foam plastics are open-cell cellular materials with the capability of absorbing oils and solvents. UF foam is non-toxic, nonflammable, and stable with respect to almost all organic solvents, light and heavy mineral oils, but is decomposed by dilute and concentrated acids and alkalis. It exhibits extraordinary aging stability, has good sound-absorbing properties, and has the lowest thermal conductivity, despite the open cells. These properties coupled with its light weight (bulk density 11 kg/m^3) and low manufacturing cost, make UF foam suitable for a wide spectrum of applications, including industrial filling materials, insulating in enclosed cavities, plant substrates (soil-free cultivation), and medical applications.

UF foam is used on-site to fill cavities of all shapes and sizes, whether of natural origin or resulting from construction wall or other applications. It has been successfully employed in mining for over 50 years. The complete filling of cavities eliminates the hazard of methane accumulation and reduces the danger of fire and explosion. UF foam filling is an inexpensive, rapid, and excellent heat-insulated lightweight construction method. When insulation is retrofitted, the foam is introduced into the existing cavities through sealable small holes.

When UF foam is formed, formaldehyde is released. It is important to make sure that the proper ratio of components is employed and suitable construction measures are taken, as otherwise the problems of formaldehyde release from foam over short term or long term may be encountered. With present day technologies, it is possible to satisfy strict conditions that a formaldehyde level of 0.1 ppm should not be exceeded in the air of a room used continuously for dwelling purposes.

Geo- and *hydroponics*, as well as *plastoponics* are terms referring to the cultivation and breeding of plants with foam as flakes or in solid form. Beans, potatoes, carrots, tomatoes and ornamental plants have been grown extensively in foam. Its suitability for land recovery in desert and semi-desert regions has been established through extensive testing.

1.17.8 Silicone Foams

Silicone foams result from the condensation reaction between ≡SiH and ≡SiOH shown below:

$$\equiv \text{SiH} + \equiv \text{SiOH} + \text{Catalyst} \rightarrow \equiv \text{Si–O–Si} \equiv + \text{H}_2 \qquad (1.3)$$

When these three components (that is, ≡SiH-containing cross-linker, ≡SiOH-containing polymer, and catalyst) are mixed together, both blowing (generation of hydrogen gas) and curing or cross-linking, that is, formation of siloxane linkage (≡Si–O–Si≡) occur. It should be noted, however, that a cross-linked product forms if the functionality of the ≡SiH-containing component is 3 or greater and the ≡SiOH-containing compound has a functionality of at least 2. These reactions are heat accelerated, but they occur readily at room temperature in the presence of catalyst. These RTV (room temperature vulcanizing) foams are thus two-pack systems. Generally, the ≡SiH (such as methylhydrogen siloxane) and catalyst make up the second component. A variety of catalysts can be used to promote the reaction. Chloroplatinic acid or other soluble platinum compound is most commonly used because it imparts flame retardancy to the formulation.

If a vinyl endblocked polymethylsiloxane is used in place of polydimethylsiloxane in the above formulation, then another competing reaction can also occur as shown below:

$$\equiv \text{SiH} + \equiv \text{SiCH} = \text{CH}_2 + \text{Catalyst} \rightarrow \equiv \text{Si–CH}_2\text{–CH}_2\text{–Si} \equiv \qquad (1.4)$$

The reaction is also catalyzed by platinum. The addition of some vinyl-containing polysiloxane can thus improve properties such as density, tensile strength, cure rate, and so forth.

Although hydrogen generation [Equation 1.3] is the most prevalent method of blowing silicone foam, there are other approaches. Adding a gas at a high enough pressure (so its volume is low before it expands) is an easy way to make foam. Gases commonly used are N_2, CO_2, and air-pressurized liquefied gases such as CFCs. One may also use chemical blowing agents (see later), which decompose generating gas when heat is applied or pH is changed. Nitrogen-liberating organic blowing agents are used extensively for foaming silicone gum.

Silicones in general are inert to most environmental agents and have many unique properties (see "Silicones" in Chapter 1 of *Industrial Polymers, Specialty Polymers, and Their Applications*). When siloxane polymers are processed into foams they carry with them most of their durability characteristics and characteristic properties. Silicone foams are thus used in a wide range of applications. Flexible foam sheet is used in airplanes as the material for fire blocking, insulation of air ducts, gasketing in engine housing compartments, and shock absorbers. Silicone foam is a popular choice in the construction industry, because of its weatherability, thermal insulation, and sealing capability.

1.17.9 Phenolic Foams

Phenolic foam is a light weight foam created from phenolic resins. It is used in a large range of applications, such as flower foam blocks, building thermal insulation, fire protection, damping, and civil engineering in a wide variety of shapes: blocks, sheets, and sprayed foams.

Phenolic foams are generally made using a resol-type phenolic resin (or a resin blend typically containing 60% of a resol-type resin and 40% of a resorcinol-modified novolac resin), surfactants, blowing agents, catalysts, and additives. Surfactants and additives are mixed into the resin and the blend is then mixed with the liquid blowing agent(s) and finally with an acid catalyst, e.g., H_3PO_4.

Surfactants are used to control cell size and structure. The most common surfactants are siloxane-oxyalkylene copolymers, polyoxyethylene sorbitan fatty acid esters, and the condensation products of ethylene oxide with castor oil and alkyl phenols. A commonly added additive is urea which is used as a formaldehyde scavenger. Very fine particle size inorganic fillers can be added to act as nucleating sites and to promote finer, more uniform cell structure, as well as increased compressive strength, but at a cost of higher density.

The most common blowing agents used for making phenolic foams are organic liquids that have boiling points approximately in the range 20°C–90°C. Suitable blowing agents include HFCs, HCFCs and others, and hydrocarbons having from about 3–10 carbon atoms such as pentane, hexane and petroleum ether. Hydrocarbons such as isopentane, isobutane and hexane are the preferred blowing agents for flower blocks. In insulating foam sector, however, the non-hydrocarbon blowing agent alternatives are not as viable and Foranil fluorochemical blowing agents have been mainly used for their insulating properties and non-flammability.

Phenolic foams can also be prepared without the use of CFC or hydrocarbon blowing agents. In a typical preparation [37], resol 200, ethoxylated castor oil 8, boric anhydride accelerator 36, and $SnCl_2.2H_2O$ 36 parts are mixed and heated for about 4 minute at 120°C to obtain a foam having density 0.003 g/cm^3 and 29% closed cells. In another method [38], 1 mol phenol, 2.6 mol formaldehyde, and 5% dimethylaminoethanol are heated at 70°C–100°C for 4 h to obtain a liquid (70%–80% solids content) which is mixed with 2%–3% silicone foam regulator and 5 parts $NaHCO_3$. Toluenesulfonic acid (20 parts, 80% aq.) is then added to obtain a stable rigid foam.

1.17.10 Poly(Vinyl Chloride) Foams

A number of methods have been devised for producing cellular products from PVC, either by a mechanical blowing process or by one of several chemical blowing techniques. PVC foams are produced in rigid or flexible forms. The greatest interest in rigid PVC foam is in applications where low-flammability requirements prevail. It has an almost completely closed cell structure and therefore low water absorption. The rigid PVC foam is used as the cellular layer of some sandwich and multi-layer panels.

Plastisols are the most widely used route to flexible expanded PVC products. Dolls, gaskets, and resilient covers for tool handles, for example, are produced from expandable plastisol compounds by molding, while varied types of upholstery, garment fabrics, and foam layer in coated-fabric flooring are made from coatings with such compounds. Figure 1.69 shows a schematic representation of the German Trovipor process for producing flexible, mainly open cell, and low to medium density (60–270 kg/m^3, 3.75–16.87 lb/ft.^{3}) PVC foam. It is normally produced in the form of continuous sheet (Figure 1.69).

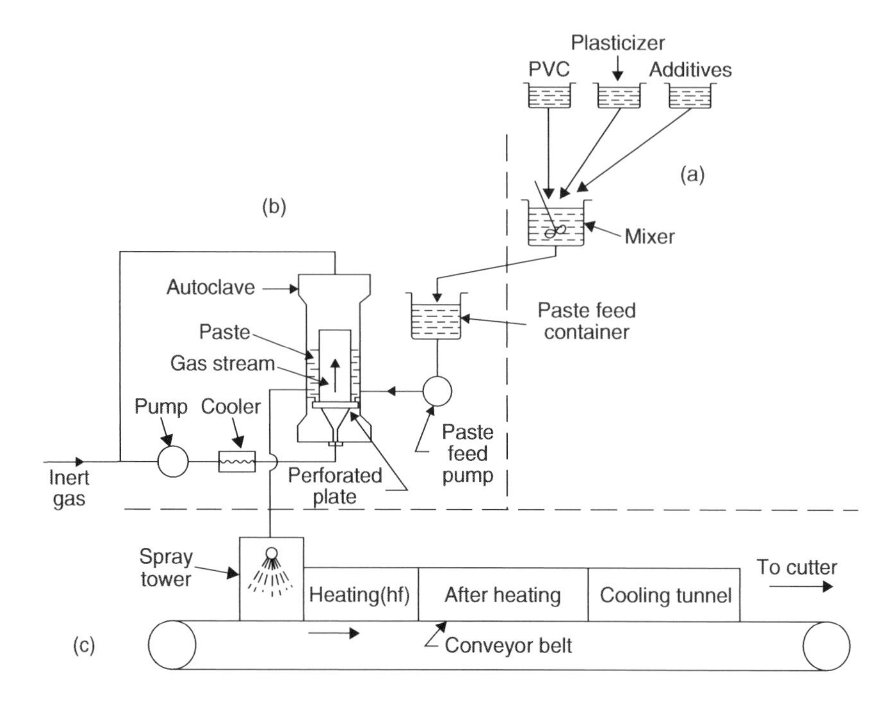

FIGURE 1.69 Schematic representation of the Trovipor process. (a) PVC paste preparation. (b) Gasification of PVC paste. (c) Spraying and fusion.

Many *chemical blowing (foaming) agents* have been developed for cellular elastomers and plastics, which, generally speaking, are organic nitrogen compounds that are stable at normal storage and mixing temperatures but undergo decomposition with gas evolution at reasonably well-defined temperatures. Three important characteristics of a chemical blowing agent are the decomposition temperature, the volume of gas generated by unit weight ("gas number," defined as the volume of gas, in cm^3, liberated by the transformation of 1 g of the blowing agent per minute), and the nature of the decomposition products. Since nitrogen is an inert, odorless, nontoxic gas, nitrogen-producing organic substances are preferred as blowing agents. Several examples of blowing agents [39] especially recommended for vinyl plastisols are shown in Table 1.4; in each case the gas generated is nitrogen.

To produce uniform cells, the blowing agent must be uniformly dispersed or dissolved in the plastisol and uniformly nucleated. It should decompose rapidly and smoothly over a narrow temperature range corresponding to the attainment of a high viscosity or gelation of the plastisol system. The gelation involves solvation of the resin in plasticizer at 300°F–400°F (149°C–204°C), the temperature depending on the ingredients employed in the plastisol. The foam quality is largely determined by the matching of the decomposition of the blowing agent to the gelation of the polymer system. If gelation occurs before gas evolution, large holes or fissures may form. On the other hand, if gas evolution occurs too soon before gelation, the cells may collapse, giving a coarse, weak, and spongy product.

Among the blowing agents listed in Table 1.4, azobisformamide (ABFA) is the most widely used for vinyls because it fulfills the requirements efficiently. ABFA decomposition can also be adjusted through proper choice of metal organic activators so that the gas evolution occurs over a narrow range within the wide range given in Table 1.4.

Though the gas number of ABFA is normally 220–260 cm^3/g, it can go up to 420 cm^3/g in the presence of catalysts. Azodicarbonamide is recommended for foaming of PVC, polyolefins, polyamides, polysiloxanes, epoxides, polymers and compolymers of acrylonitrile and acrylates, and rubbers.

TABLE 1.4 Commercial Blowing (Foaming) Agents

Chemical Type	Decomposition Temperature in Air (°C)	Decomposition Position Range in Plastics (°C)	Gas Yield (mL/g)
Azo compounds			
Azobisformamide (Azodicarbonamide) (ABFA)	195–200	160–200	220
Azobisisobutyronitrile (AIBN)	115	90–115	130
Diazoaminobenzene (DAB)	103	95–100	115
N-Nitroso compounds			
N, *N*′-Dimethyl-*N*, *N*′-dinitrosoterephthalimide (DTA)	105	90–105	126
N, *N*′-Dinitrosopentamethylenetetramine (DNPA)	195	130–190	265
Sulfonyl hydrazides			
Benzenesulfonylhydrazide (BSH)	>95	95–100	130
Toluene-(4)-sulfonyl hydrazide (TSH)	103	100–106	120
Benzene-1, 3-disulfonyl hydrazide (BDH)	146	115–130	85
4, 4′-oxybis(benzenesulfonylhydrazide) (OBSH)	150	120–140	125

Source: Lasman, H. R. 1967. *Mod Plastics*, 45, 1A, Encycl. Issue, 368.

Diazoaminobenzene (DAB) is one of the first organic blowing agents to find industrial application. Its decomposition point (95°C–150°C) and gas number (115 cm^3/g) depend on the pH of the medium; in acidic media it decomposes at lower temperature and more completely. DAB is used in foaming phenolic and epoxy resins, PVC, rubber and other high poymers.

N,N′-Dinitrosopentamethylenetetramine (DNPA) is the cheapest (except for urea oxalate) and most widely used organic blowing agent accounting for 50% of all blowing agents used. It however disperses poorly in mixtures and is sensitive to shock and friction (explosive).

Because of the relatively low temperature of decomposition, DTA can be used to make foams with a uniform cellular structure without deterioration of the polymer. The disadvantages of DTA are, however, poor dispersive ability in mixtures and sensitivity to moisture. Nevertheless, DTA is used in foaming PVC (especially for thin walled articles), polyurethane, polystyrene, polyamides, and siloxane rubbers.

BSH is used for foaming rubbers, polystyrene, epoxy resins, polyamides, PVC, polyesters, phenol-formaldehyde resins, and polyolefins. However, the thermal decomposition of BSH yields not only nitrogen but also a nontoxic residue (disulfide and thiosulfone) which may degrade to give thiophenol and thus an unpleasant odor to the foams.

OBSH is one of the best blowing agents of the sulfonylhydrazide class. Its gas liberation characteristics (no stepwise change up to 140°C) makes it possible to obtain foams with small, uniform cells. It is nontoxic and does not impart color and smell to articles. OBSH is used for foaming PVC, polyolefins, polysulfides, microporous rubber, or foamed materials based on mixtures of polymers with rubbers. In the last case, OBSH acts additionally as a cross-linking agent.

Closed-cell foams result when the decomposition and gelation are carried out in a closed mold almost filled with plastisol. After the heating cycle, the material is cooled in the mold under pressure until it is dimensionally stable. The mold is then opened, and the free article is again subjected to heat (below the previous molding temperature) for final expansion. Protective padding, life jackets, buoys, and floats are some items made by this process.

The blowing agents given in Table 1.4 can be used to make foamed rubber. A stable network in this product results from the cross-linking reaction (vulcanization), which thus corresponds to the step of fusion in the case of plastisols. Some thermoplastics also can be foamed by thermal decomposition of blowing agents even though they do not undergo an increase in dimensional stability at an elevated temperature. In this case the viscosity of the melt is high enough to slow down the collapse of gas bubbles so that when the polymer is cooled below its T_m a reasonably uniform cell structure can be built in. Cellular polyethylene is made in this way.

1.17.11 Special Foams

Some special types of foams are: (1) structural foams; (2) syntactic foams and multifoams; and (3) reinforced foams. Structural foams (Figure 1.58c and d), which possess full-density skins and cellular cores, are similar to structural sandwich constructions or to human bones, which have solid surfaces but cellular cores. Structural foams may be manufactured by high pressure processes or by low-pressure processes (Figure 1.61). The first one may provide denser, smoother skins with greater fidelity to fine detail in the mold than may be true of low-pressure processes. Fine wood detail, for example, is used for simulated wood furniture and simulated wood beams. Surfaces made by low-pressure processes may, however, show swirl or other textures, not necessarily detracting from their usefulness. Almost any thermoplastic or thermosetting polymer can be formulated into a structural foam.

In the case of syntactic foams (or spheroplastics), instead of employing a blowing agent to form bubbles in the polymer mass, hollow spherical particles, called microspheres, microcapsules, or microballoons, are embedded in a matrix of unblown polymer. (In *multifoams*, microspheres are combined with a foamed polymer to provide both kinds of cells.) Since the polymer matrix is not foamed, but is filled mechanically with the hollow spheres, syntactic materials may also be thought of as reinforced or filled plastics, with the gas-containing particles being the reinforcing component. Synthetic wood, for instance, is provided by a mixture of polyester and small hollow glass spheres (microspheres).

The cellular structure of the syntactic foam depends on the size, quantity, and distributive uniformity of the microspheres. Since the microspheres have continuous shells, the final material will, as a rule, have completely enclosed cells, and thus can be called absolute foamed plastics or "absolute" closed-cell foams. This, together with the absence of microstructural anisotropy (because the microspheres have practically all the same size and are uniformly distributed in the matrix), gives a syntactic material its valuable properties. They have better strength-to-weight ratios than conventional foamed plastics, absorb less water, and can withstand considerable hydrostatic pressures. Using hollow sphere means that the final material is lighter than one containing a compact filler, such as glass powder, talc, kaolin, quartz meal, or asbestos.

Figure 1.70a and b are graphical presentations of syntactic foam structure in which the two components, microspheres and resin fill completely the whole volume (no dispersed voids) and the density of the product is thus calculated from the relative proportion of the two. Measured density values often differ from the calculated ones due to the existence of some isolated or interconnected irregularly shaped voids, as shown in Figure 1.70c. The voids are usually an incidental part of the composite, as it is not easy to avoid their formation. Nevertheless, voids are often introduced intentionally to reduce the density below the minimum possible in a close-packed two-phase structure (Figure 1.70a).

Syntactic foams exhibit their best mechanical behavior in the compressive mode. The spheres themselves are an extremely strong structure and hence can withstand such stresses very well. Syntactic materials consisting of hollow glass microspheres in epoxy resin are used for sandwich structures and as potting compounds for high-density electronic modules and other units likely to encounter hydrostatic pressures. Hollow glass microspheres and powdered aluminum in resin are used as core materials for sandwich construction and radomes. Hollow glass microspheres in aluminum matrix are used for aerospace and extreme hydrostatic pressure (oceanographic) applications in view of low weight and high compressive strength.

Polymer foams may be reinforced, usually with short glass fibers, and also other fibers such as asbestos or metal, and other reinforcements such as carbon black. The reinforcing agent is generally introduced into the basic components and is blown along with them, to form part of and to reinforce the walls of the cells (Figure 1.58e). When this is done, it is not unusual to obtain increases in mechanical properties of 400–500% with fiberglass content up to 50% by weight, especially in thermosettings. The principal advantages of reinforcement, in addition to increased strength and stiffness, are improved dimensional stability, resistance to extremes of temperature and resistance to creep.

Two processes for the manufacture of glass-reinforced foam laminates used as building materials, namely, free-rise process and restrained rise process are presented in Figure 1.71 and Figure 1.72. The glass fiber reinforcement is a thin (0.25–1.25 mm) mat supplied in roll form. It consists of layers of relatively long (1.5–4 m) glass fibers, the fibers in one layer being at an acute angle to the fibers in each next adjacent layer. A small amount of silane-modified polyester, or other binder is present, at a level of 2%–10% by weight. This type of glass mat is relatively porous to the passage of liquids and is also capable of expanding within a mixture of rising foam chemicals to provide a uniform, three-dimensional

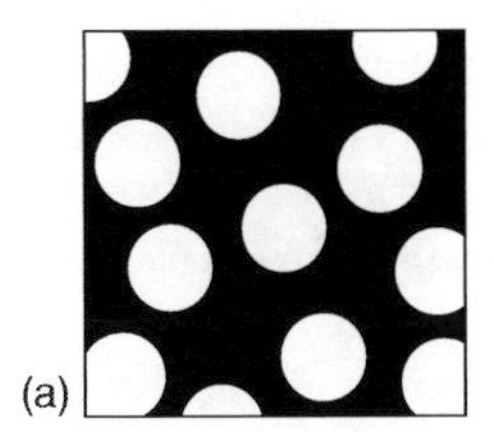

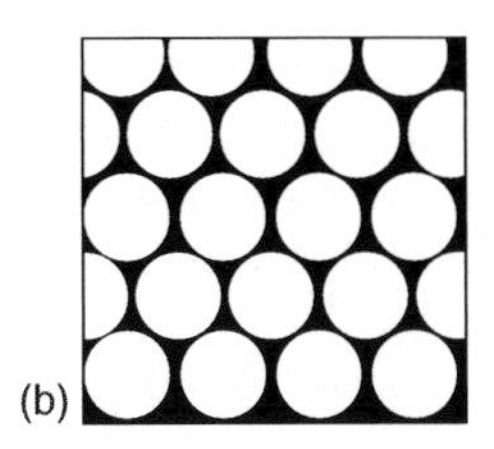

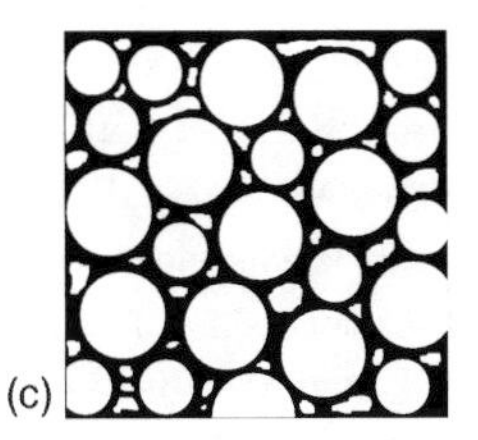

FIGURE 1.70 Graphical representations of syntactic foam structures. (a) Two-phase composite with random dispersion of spheres. (b) Two-phase composite with hexagonal close-packed structure of uniform sized spheres (74% by vol.). (c) Three-phase composite containing packed microspheres, dispersed voids and binding resin.

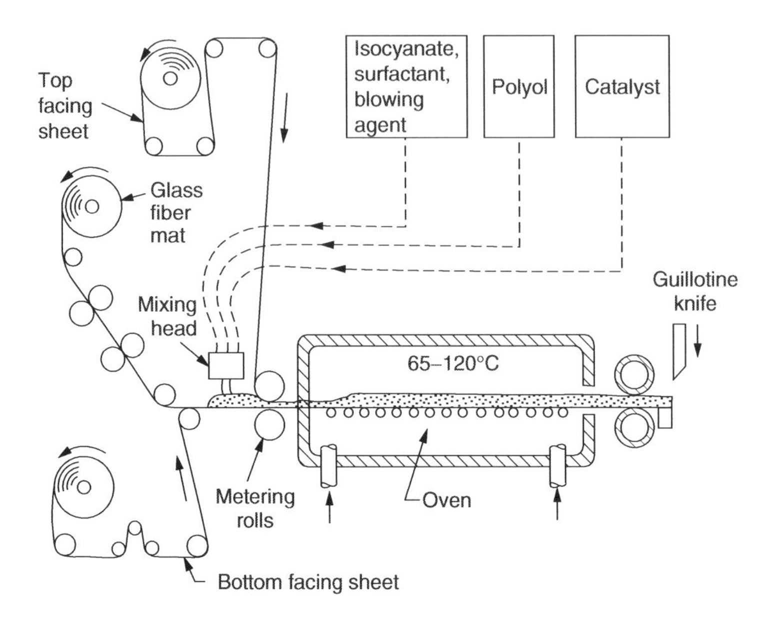

FIGURE 1.71 Schematic of a free-rise process for manufacture of glass-reinforced foam laminates.

reinforcing network within the final foamed laminate. The glass fiber reinforcement is functionally effective when used at levels of 4–24 g per board foot of the laminate [40]. Various types of facing sheets may be used, such as aluminum foil for building insulation products, asphalt-saturated felts for roof insulation or any other material, e.g., paper and plastic films.

1.18 Rubber Compounding and Processing Technology

1.18.1 Compounding Ingredients

No rubber becomes technically useful if its molecules are not cross-linked, at least partially, by a process known as curing or vulcanization [41–44]. For NR and many synthetic rubbers, particularly the diene rubbers, the curing agent most commonly used is sulfur. But sulfur curing takes place at technically viable rates only at a relatively high temperature (>140°C) and, moreover, if sulfur alone is used, optimum curing requires use of a fairly high dose of sulfur, typically 8–10 parts per hundred parts of rubber (phr), and heating for nearly 8 h at 140°C.

Sulfur dose has been substantially lowered, however, with the advent of organic accelerators. Thus, incorporation of only 0.2–2.0 phr of accelerator allows reduction of sulfur dose from 8–10 to 0.5–3 phr and effective curing is achieved in a time scale of a few minutes to nearly an hour depending on temperature (100°C–140°C) and type of the selected accelerator. The low sulfur dose required in the accelerated sulfur vulcanization has not only eliminated bloom (migration of unreacted sulfur to the surface of the vulcanizate), which was a common feature of the earlier nonaccelerated technology, but also has led to the production of vulcanizates of greatly improved physical properties and good resistance to heat and aging.

The selection of the accelerator depends largely on the nature of the rubber taken, the design of the product, and the processing conditions. It is important to adopt a vulcanizing system that not only gives a rapid and effective cross-linking at the desired vulcanizing temperatures but also resists premature vulcanization (scorching) at somewhat lower temperatures encountered in such operations as mixing, extrusion, calendaring, and otherwise shaping the rubber before final cross-linking. This may require the use of delayed-action type accelerators as exemplified by sulfenamides. Other principal types of accelerators with different properties are guanidines, thiazoles, dithiocarbamates, thiurams, and

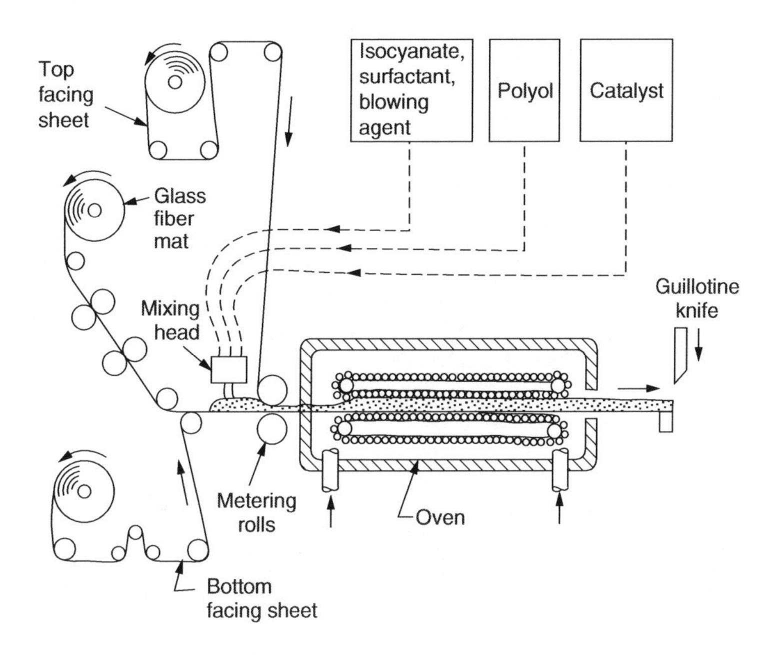

FIGURE 1.72 Schematic of a restrained rise process for manufacture of glass-reinforced foam laminates.

xanthates (Table 1.5). Accelerators are more appropriately classified according to the speed of curing induced in their presence in NR systems. In the order of increasing speed of curing, they are classified as slow, medium, semiultra, and ultra accelerators.

The problem of scorching or premature vulcanization is very acute with ultra or fast accelerators. Rubber stocks are usually bad conductors of heat and therefore flow of heat to the interior of a vulcanizing stock from outside is very slow. As a result, in thick items the outer layers may reach a state of overcuring before the core or interior layers begin to cure. For such thick items, a slow accelerator (Table 1.5) is most suitable.

For butyl and EPDM rubbers, which have very limited unsaturations, slow accelerators are, however, unsuitable and, fast accelerators should be used at high temperatures for good curing at convenient rates. Since butyl rubber is characterized by *reversion*, a phenomenon of decrease of tensile strength and modulus with time of cure after reaching a maximum, duration of heating at curing temperatures must be carefully controlled, and prolonged heating must be avoided.

For rubbers with higher degree of unsaturations, an ideal accelerator is one that is stable during mixing, processing, and storage of the mix, but that reacts and decomposes sharply at the high vulcanization temperature to effect fast curring. These requirements or demands are closely fulfilled by the delayed-action accelerators typical examples of which are given in Table 1.5.

The spectacular effects of modern organic accelerators in sulfur vulcanization of rubber are observed only in the presence of some other additives known as accelerator activators. They are usually two-component systems comprising a metal oxide and a fatty acid. The primary requirement for satisfactory activation of accelerator is good dispersibility or solubility of the activators in rubber. Oxides of bivalent metals such as zinc, calcium, magnesium, lead, and cadmium act as activators in combination with stearic acid. A combination of zinc oxide and stearic acid is almost universally used. (Where a high degree of transparency is required, the activator may be a fatty acid salt such as zinc stearate.) Besides speeding up the rate of curing, activators also bring about improvements in physical properties of vulcanizates. This is highlighted by the data given in Table 1.6.

A sulfur-curing system thus has basically four components: a sulfur vulcanizing agent, an accelerator (sometimes combinations of accelerators), a metal oxide, and a fatty acid. In addition, in order to improve

TABLE 1.5 Accelerator for Sulfur Vulcanization of Rubbers

Accelerator Type and Formula	Chemical Name	Accelerator Activity
Guanidines		
$(C_6H_5NH)_2C=NH$	Diphenyl guanidine (DPG)	Medium accelerator
$(C_6H_5NH)_2C=N-C_6H_5$	Triphenyl guanidine (TPG)	Slow accelerator
Thiazoles		
Benzothiazole C–SH	Mercaptobenzothiazole (MBT)	Semi-ultra accelerator
Benzothiazole C–S–S–C benzothiazole	Mercaptobenzothiazyl disulfide (MBTS)[a]	Semi-ultra (delayed action) accelerator

Structure	Name	Function
Sulfenamides		
Benzothiazole (N, S)–C–S–NH–cyclohexyl	*N*-Cyclohexyl benzothiazyl sulfenamide (CBS)	Semi-ultra (delayed action) accelerator
Benzothiazole (N, S)–C–S–N(morpholine, O)	*N*-Oxydiethylenebenzothiazyl sulfenamide (NOBS) or 2-Morpholinothiobenzothiazole (MBS)	Semi-ultra (delayed action) accelerator
Benzothiazole (N, S)–C–S–NH–C_4H_9	*N*-t-Butylbenzothiazyl sulfenamide (TBBS)	Semi-ultra (delayed action) accelerator
Dithiocarbamates		
$\left((C_2H_5)_2N-C(=S)-S^-\right)_2 Zn^{2+}$	Zinc diethyl Dithiocarbamate (ZDC)	Ultra accelerator
$(C_2H_5)_2N-C(=S)-S^-\ Na^+$	Sodium diethyl dithiocarbamate (SDC)	Ultra accelerator, water soluble (used for latex)

(continued)

Table 1.5 (Continued)

Accelerator Type and Formula	Chemical Name	Accelerator Activity
Thiuram sulfides		
$(CH_3)_2N-C(=S)-S-S-C(=S)-N(CH_3)_2$	Tetramethyl thiuram disulfide (TMTD, TMT)[a]	Ultra accelerator
$(C_2H_5)_2N-C(=S)-S-S-C(=S)-N(C_2H_5)_2$	Tetraethyl thiuram disulfide (TETD, TET)[a]	Ultra accelerator
$(CH_3)_2N-C(=S)-S-C(=S)-N(CH_3)_2$	Tetramethyl thiuram monosulfide (TMTM)	Ultra accelerator

Xanthates		
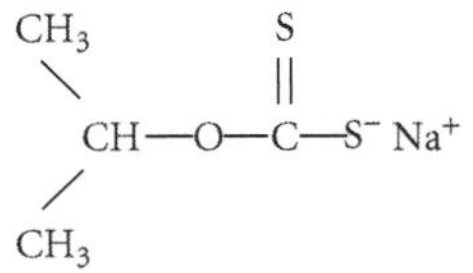 $(CH_3)_2CH{-}O{-}C(=S){-}S^- \; Na^+$	Sodium isopropyl xanthate (SIX)	Ultra accelerator, water soluble (suited for latex)
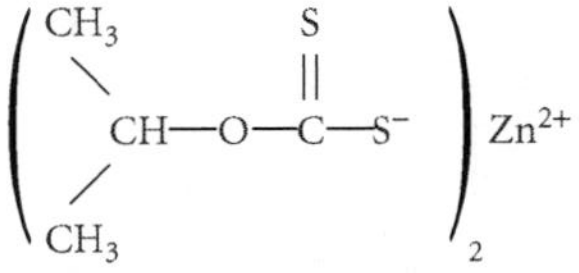 $\left((CH_3)_2CH{-}O{-}C(=S){-}S^-\right)_2 Zn^{2+}$	Zinc isopropyl xanthate (ZIX)	Ultra accelerator

[a] Sulfur donors.

TABLE 1.6 Effect of Activator on Vulcanization

	Tensile Strength, Psi (MPa)	
	ZnO (phr)	
Time of Cure (min)	0.0	5.0
	100 (0.7)	2300 (16)
30	400 (2.8)	2900 (20)
60	1050 (7.2)	2900 (20)
90	1300 (9.0)	2900 (20)

Base compound: NR (pale creep) 100; sulfur 3; mercaptobenzothiazole (MBT) 0.5. Temperature of vulcanization 142°C.

the resistance to scorching, a prevulcanization inhibitor such as *N*-cyclohexylthiophthalimide may be incorporated without adverse effects on either the rate of cure or physical properties of the vulcanizate.

The level of accelerator used varies from polymer to polymer. Some typical curing systems for the diene rubbers (NR, SBR, and NBR) and for two olefin rubbers (IIR and EPDM—see Appendix A2 of *Industrial Polymers, Specialty Polymers, and Their Applications* for abbreviations) are given in Table 1.7.

In addition to the components of the vulcanization system, several other additives are commonly used with diene rubbers. Rubbers in general, and diene rubbers in particular, are blended with many more additives than is common for most thermoplastics with the possible exceptions of PVC. The major additional classes of additives are:

1. Antidegradants (antioxidants and antiozonants)
2. Processing aids (peptizers, plasticizers, softeners, and extenders, tackifiers, etc.)
3. Fillers
4. Pigments
5. Others (retarders, blowing agents)

The use of antioxidants and antiozonants has already been described in Chapter 1 of *Plastics Fundamentals, Properties, and Testing.*

1.18.1.1 Processing Aids

Peptizers are added to rubber at the beginning of mastication (see later) and are used to increase the efficiency of mastication. They act chemically and effectively at temperatures greater than 65°C and hasten the rate of breakdown of rubber chains during mastication. Common peptizers are zinc thiobenzoate, zinc-2-benzamidothiophenate, thio-β-naphthol, etc. Processing aids other than the peptizer and compounding ingredients (additives) are added after the rubber attains the desired plasticity on mastication.

Common process aids, besides the peptizer, are pine tar, mineral oil, wax, factice, coumarone-indene resins, petroleum resins, rosin derivatives, and polyterpenes. Their main effect is to make rubber soft and

TABLE 1.7 Components of Sulfur Vulcanization Systems

	Rubber				
Additive[a] (phr)	NR	SBR	NBR	IIR	EPDM
Sulfur	2.5	2.0	1.5	2.0	1.5
Zinc oxide	5.0	5.0	5.0	3.0	5.0
Stearic acid	2.0	2.0	1.0	2.0	1.0
TBBS	0.6	1.0	—	—	—
MBTS	—	—	1.0	0.5	—
MBT	—	—	—	—	1.5
TMTD	—	—	0.1	1.0	0.5

[a] See Table 1.5 for accelerator abbreviations.

tacky to facilitate uniform mixing, particularly when high loading of carbon black or other fillers is to be used.

Factice (vulcanized oil) is a soft material made by treating drying or semidrying vegetable oils with sulfur monochloride (cold or white factice) or by heating the oils with sulfur at 140–160°C (hot or brown factice). The use of factice (5–30 phr) allows efficient mixing and dispersion of powdery ingredients and gives a better rubber mix for the purpose of extrusion.

Ester plasticizers (phthalates and phosphates) that are used to plasticize PVC (see Chapter 1 of *Plastics Fundamentals, Properties, and Testing*) are also used as process aids, particularly with NBR and CR. Polymerizable plasticizers such as ethylene glycol dimethacrylate are particularly useful for peroxide curing rubbers. They act as plasticizers or tackifiers during mixing and undergo polymerization by peroxide initiation during cure.

The diene hydrocarbon rubbers are often blended with hydrocarbon oils. The oils decrease polymer viscosity and reduce hardness and low temperature brittle point of the cured product. They are thus closely analogous to the plasticizers used with thermoplastics but are generally known as softners. Three main types of softners are distinguished: aliphatic, aromatic, and naphthenic. The naphthenics are preferred for general all-round properties.

NRs exhibit the phenomenon known as tack. Thus when two clean surfaces of masticated rubber are brought into contact the two surfaces strongly adhere to each other, which is a consequence of interpenetration of molecular ends followed by crystallization. Amorphous rubbers such as SBR do not display such tack and it is necessary to add tackifiers such as rosin derivatives and polyterpenes.

1.18.1.2 Fillers

The principles of use of inert fillers, pigments, and blowing agents generally follow those described in Chapter 1 of *Plastics Fundamentals, Properties, and Testing.* Major fillers used in the rubber industry are classified as (1) nonblack fillers such as china clay, whiting, magnesium carbonate, hydrated alumina, anhydrous, and hydrated silicas and silicates including those in the form of ground mineral such as slate powder, talc, or French chalk, and (2) carbon blacks.

Rather peculiar to the rubber industry is the use of the fine particle size reinforcing fillers, particularly carbon black. Fillers may be used from 50 phr to as high as 100–120 phr or even higher proportions. Their use improves such properties as modulus, tear strength, abrasion resistance, and hardness. They are essential with amorphous rubbers such as SBR and polybutadiene that has little strength without them. They are less essential with strain-crystallizing rubbers such as NR for many applications but are important in the manufacture of tires and related products.

Carbon blacks are essentially elemental carbon and are produced by thermal decomposition or partial combustion of liquid or gaseous hydrocarbons to carbon and hydrogen. The principal types, according to their method of production, are channel black, furnace black, and thermal black.

Thermal black is made from natural gas by the thermatomic process in which methane is cracked over hot bricks at a temperature of 1,600°F (871°C) to form amorphous carbon and hydrogen. Thermal black consists of relatively coarse particles and is used principally as a pigment. A few grades (FT and MT referring to fine thermal and medium thermal) are also used in the rubber industry.

Most of the carbon black used in the rubber industry is made by the furnace process (furnace black), that is, by burning natural gas or vaporized aromatic hydrocarbon oil in a closed furnace with about 50% of the air required for complete combustion. Furnace black produced from natural gas has an intermediate particle size, while that produced from oil can be made in wide range of controlled particle sizes and is particularly suitable for reinforcing rubbers.

Quite a variety of grades of furnace blacks are available, e.g., fine furnace black (FF), high modulus (HMF), high elongation (HEF), reinforcing (RF), semireinforcing (SRF), high abrasion (HAF), super abrasion (SAF), intermediate super abrasion (ISAF), fast extruding (FEF), general purpose (GPF), easy processing (EPF), conducting (CF), and super conducting furnace black (SCF).

Channel black is characterized by lower pH, higher volatile content, and high surface area. It has the smallest particle size of any industrial material. A few grades of channel blacks (HPC, MPC, or EPC corresponding to hard, medium, or easy processing channel) are used in the rubber industry.

For carbon-black fillers, structure, particle size, particle porosity, and overall physico-chemical nature of particle surface are important factors in deciding cure rate and degree of reinforcement attainable. The pH of the carbon black has a profound influence. Acidic blacks (channel blacks) tend to retard the curing process while alkaline blacks (furnace blacks) produce a rate-enhancing effect in relation to curing, and may even give rise to scorching.

Another important factor is the particle size of the carbon black filler. The smaller the particle size, the higher the reinforcement, but the poorer the processability because of the longer time needed for dispersion and the greater heat produced during mixing. Blacks of the smallest particle size are thus unsuitable for use in rubber compounding.

For carbon black fillers the term *structure* is used to represent the clustering together and entanglement of fine carbon particles into long chains and three-dimensional aggregates. High-structure blacks produce high-modulus vulcanizates as high shear forces applied during mixing break the agglomerates down to many active free radical sites, which bind the rubber molecules, thereby leading to greater reinforcement. In nonstructure blacks the aggregates are almost nonexistent.

Most of the nonblack fillers used in rubber compounds are of nonreinforcing types. They are added for various objectives, the most important being cost reduction. Precipitated silica (hydrated), containing about 10–12% water with average particle size ranging 10–40 nm, produce effective reinforcements and are widely used in translucent and colored products. Finely ground magnesium carbonate and aluminum silicate also induce good reinforcing effects. Precipitated calcium carbonate and activated calcium carbonate (obtained by treating calcium carbonate with a stearate) are used as semireinforcing fillers.

Short fibers of cotton, rayon, or nylon may be added to rubber to enhance modulus and tear and abrasion resistance of the vulcanizates. Some resins such as "high styrene resins" and novolac-type phenolic resins mixed with hexamethylene tetramine may also be used as reinforcing fillers or additives. Whereas SBR has a styrene content of about 23.5% and is rubbery, styrene-butadiene copolymers containing about 50% styrene are leatherlike whilst with 70% styrene the materials are more rigid thermo-plastics but with low softening points. Both of these copolymers are known in the rubber industry as high styrene resins and are usually blended with a hydrocarbon rubber such as NR and SBR. Such blends have found use in shoe soles, car wash brushes and other moldings, but in recent years have suffered increasing competition from conventional thermoplastics and thermoplastic rubbers.

1.18.2 Mastication and Mixing

A deficiency of NR, compared with the synthetics, is its very high molecular weight, which makes mixing of compounding ingredients and subsequent processing by extrusion and other shaping operations difficult. For NR it is thus absolutely necessary, while for synthetic rubbers it is helpful, to subject the stock to a process of breakdown of the molecular chains prior to compounding. This is effected by subjecting the rubber to high mechanical work (shearing action), a process commonly known as *mastication.* Mastication and mixing are conveniently done using two-roll mills and internal mixers. The oxygen in air plays a critical role during mastication.

Rubber that has been masticated is more soft and flows more readily than the unmasticated material. Mastication also allows preparation of solutions of high solids content because of the much lower solution viscosity of the degraded rubber. Rubber is also rendered tacky by mastication, which means that the uncured rubber sticks to itself readily so that articles of suitable thickness can be built up from layers of masticated rubber or rubberized fabric without the use of a solvent.

1.18.2.1 Open Mill

The mainstays of the rubber industry for over 70 years has been the two-roll (open) mill and the Banbury (internal) mixer. Roll mills were first used for rubber mixing over 120 years ago. The plastics and adhesives industries later adopted these tools.

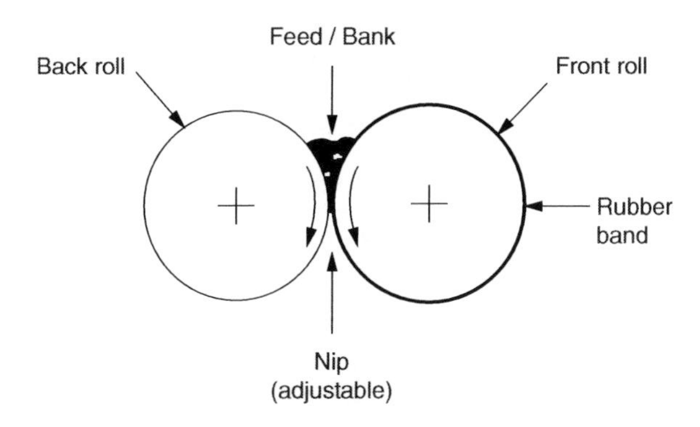

FIGURE 1.73 Section showing the features of a two-roll open mill.

The two-roll mill (Figure 1.73) consists of two opposite-rotating rollers placed close to one another with the roll axes parallel and horizontal, so that relatively small gap or nip (adjustable) between the cylindrical surface exists. The speeds of the two rolls are usually different, the front roll having a slower speed. For NR mixing, a friction ratio of 1:1.2 for the front to back roll may be used. For some synthetic rubbers or highly filled NR mixes, friction ratios close to 1.0 produce good results.

The nip is adjusted so that when pieces of rubber are placed between the rolls, they are deformed by shearing action and squeezed through the nip to which they are returned by the operator. On repeated passage through the nip, they form a band around the front roll and a moving "bank" above the nip. As the rolls keep on rotating, the operator uses a knife to cut through the band on the front roll, removes the mass in parts from time to time, and places it in a new position to ensure uniform treatment and mixing through the nip.

With most rubbers other than NR, addition of different compounding ingredients may be normally started soon after a uniform band is formed on the roll and a bank is obtained. For NR, however, milling is usually continued to masticate the rubber to the desired plasticity, and mixing of compounding ingredients is started only after the adequate mastication. Mixing is effected by adding the different ingredients onto the bank. They are gradually dispersed into the rubber, which is cut at intervals, rolled over, and recycled through the nip of the moving rolls to produce uniform mixing.

Since the rate of mastication is a function of temperature, time and temperature of mastication have to be controlled or kept uniform from batch to batch in order to get the desired uniform products from different batches.

Roll mills vary greatly in size from very small laboratory machines with rollers of about 1 in. in diameter and driven by fractional horsepower motors to very large mills with rollers of nearly 3 ft. in diameter and 7 or 8 ft. in length and driven by motors over 100 hp.

1.18.2.2 Internal Batch Mixers

Internal batch mixers are widely used in the rubber industry. They are also used for processing plastics such as vinyl, polyolefins, ABS, and polystyrene, along with thermosets such as melamines and ureas because they can hold materials at a constant temperature.

The principle of internal batch mixing was first introduced in 1916 with the development of the Banbury mixer (Figure 1.74a). A Banbury-type internal mixer essentially consists of a cylindrical chamber or shell within which materials to be mixed are deformed by rotating blades or rotors with protrusions. The mixer is provided with a feed door and hopper at the top and a discharge door at the bottom. As the rubber or mix is worked and sheared between the two rotors and between each rotor and the body of the casing, mastication takes place over the wide area, unlike in an open mill where it is restricted only in the area of the nip between the two rolls.

The rotor blade of the Banbury mixer is pear shaped, but the projection is spiral along the axis and the two spirals interlock and rotate in opposite directions (Figure 1.74b). The interaction of rotor blades between themselves, in addition to producing shearing action, causes folding or "shuffling" of the mass,

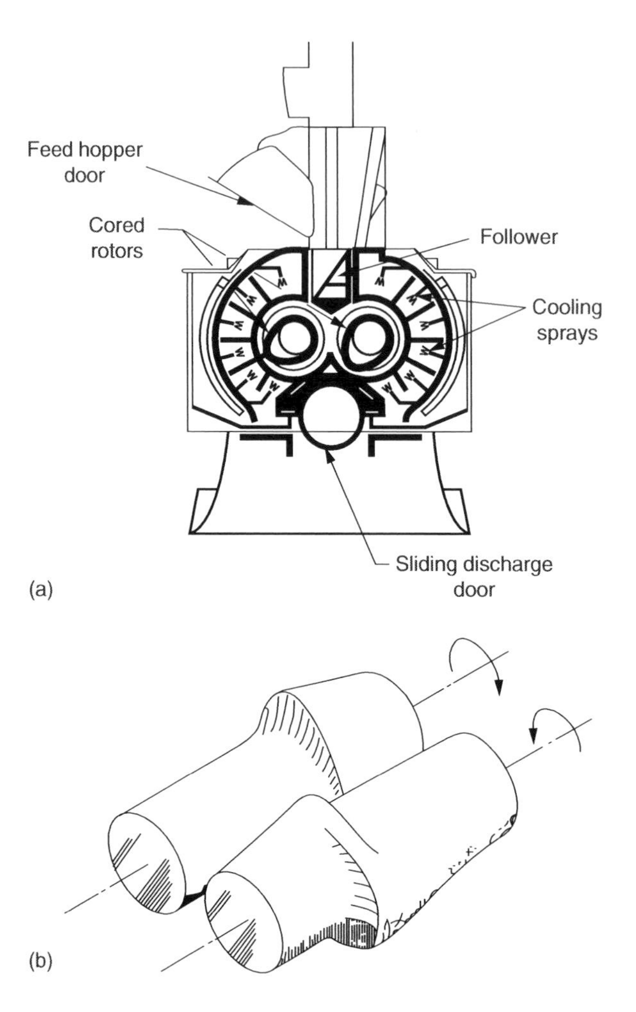

FIGURE 1.74 (a) Cross-section of a Banbury mixer; (b) Roll mixing blades in Banbury mixer. (*Farrel Co.*)

which is further accentuated by the helical arrangement of the blade along the axis of the rotor, thereby imparting motion to the mass in the third, or axial, direction. This combination of intensive working produces a highly homogeneous mix.

An important and novel feature of the Banbury mixer is a vertical ram to press the mass into contact with the two rotors. Rubbers, fillers, and other ingredients are charged through the feed hopper and then held in the mixing chamber under the pressure of the hydraulic (or manual) ram. As a result, incorporation of solids is more rapid. Both the cored rotors and the walls of the mixing chamber can be cooled or heated by circulating fluid. Because of the large power consumption of such a machine (up to 500 hp) the cylinder walls are usually water-cooled by sprays (Figure 1.74a).

Other major machines in use in the rubber industry are the Shaw intermix and the Baker-Perkins shear-mix. Both the Banbury-type and the intermix mixers have passed through modifications and refinements at different points in time. Mechanical feeding and direct oil injection in measured doses into the mixing chamber through a separate oil injection port are notable features of modern internal mixers. Higher rotor speeds and, in the Banbury type, the higher ram pressures, are used for speedy output.

In Banbury-type mixers, the rotors run at different speeds, while in intermix mixers the rotor speeds are equal but the kneading action between the thicker portion of one rotor and the thinner portion of the other produces a frictional effect.

The operation of internal mixers is power intensive, and a given job is performed at a much higher speed and in a much shorter time on a two-roll open mill. However, the major vulcanizing agent (such as sulfur) is often added later on a two-roll mill so as to eliminate possibilities of scorching. And even if this is not practiced, the mix, after being discharged from the internal mixer, is usually passed through a two-roll mill in order to convert it from irregular lumps to a sheet form for convenience in subsequent processing.

1.18.3 Reclaimed Rubber

The use of reclaimed rubber in a fresh rubber mix not only amounts to waste utilization but offers some processing and economic advantages to make it highly valued in rubber compounding. Though waste vulcanized rubber is normally not processable, application of heat and chemical agents to ground vulcanized waste rubber leads to substantial depolymerization whereby conversion of the rubber to a soft, plastic processable state is effected. Rubber so regenerated for reuse is commonly known as reclaimed rubber or simply as reclaim. Reclaimed rubber can be easily revulcanized.

Worn-out tires and scraps and trimmings of other vulcanized products constitute the raw material for reclaimed rubber. Therefore a good reclaiming process must not only turn the rubber soft and plastic but also must remove reinforcing cords and fabrics that may be present. There are a number of commercial processes [43] for rubber regeneration: (1) alkali digestion process, (2) neutral or zinc chloride digestion process, (3) heater or pan process, and (4) reclaimator process.

Tires are most commonly reclaimed by digestion processes. For processes (1) and (2), debeaded tires and scraps, cut into pieces, are ground with two-roll mills or other devices developed for the purpose. Two-roll mills generally used for grinding tires turn at a ratio of about 1:3, thus providing the shearing action necessary to rip the tire apart. The rubber chunks are screened, and the larger material is recycled until the desired size is reached. The ground rubber is then mixed with a peptizer, softener, and heavy naphtha, and charged into spherical autoclaves with requisite quantities of water containing caustic soda for process (1) or zinc chloride for process (2).

The textile is destroyed and mostly lost in the digestion process. Steam pressure and also the amount of air or oxygen in the autoclave greatly influence the period necessary for rubber reclaiming. On completion of the process, the pressure is released, the contents of the autoclave discharged into water, centrifuged, pressed to squeeze out water, and dried. The material is finally processed through a two-roll mill during which mineral fillers and oils may be added to give a product of required specific gravity and oil extension.

Butyl and NR tubes and other fiber-free scrap rubbers are reclaimed by means of the heater or pan process. Brass tube fittings and other metal are removed from the scrap. The scrap is mechanically ground, mixed with reclaiming agents, loaded into pans or devulcanizing boats, and autoclaved at steam pressures of 10–14 atm (1.03–1.40 MPa) for 3–8 h. The reclaim is finally processed much the same way as in the digester process.

The reclaimator process is more attractive than the above processes. The reclaimator is essentially a high-pressure extruder that devulcanizes fiber-free rubber continuously. Ground scrap is mechanically treated in hammer mills to remove the textile material, mixed with reclaiming oils and other materials, and then fed into the reclaimator. High pressure and shear between the rubber mixture and the extruder barrel walls effectively reclaim the rubber mixture. Devulcanization occurs at 175–205°C in a few minutes and turns the rubber into reclaim that issues from the machine continuously. The whole regeneration, which is a dry process, may be completed in about 30 min.

The reclaiming oils and chemicals are complex wood and petroleum derivatives that swell the rubber and provide access for breaking the rubber bonds with heat, pressure, chemicals, and mechanical shearing. Approximately 2–4 parts of oil are used per 100 parts of scrap rubber. Some examples of reclaiming oils include monocyclic and mixed terpenes, i.e., pine-tar products, saturated polymerized petroleum hydrocarbons, aryl disulfides in petroleum oil, cycloparafinic hydrocarbons, and alkyl aryl polyether alcohols.

Reclaimed rubber contains all the fillers present in the original scrap or waste rubber. It shows very good aging characteristics and is characterized by less heat development during mixing and processing as compared to fresh rubber. The use of reclaim in a fresh rubber mix is advantageous not on consideration of physical and mechanical properties but essentially for smooth processing and reduced cost.

Radial tires (see later) do not use reclaimed rubber because they require higher abrasion resistance that cannot be attained by mixing reclaimed rubber. Better processes for the production of higher quality reclaimed rubber are needed in order to use it for radial tires. To improve the quality of reclaimed rubber, cross-links in vulcanizates should be severed selectively during a devulcanization process and no low-molecular-weight compound such as swelling solvent should remain in the reclaimed rubber after the devulcanization process.

A devulcanization process that utilizes supercritical CO_2 as a devulcanization reaction medium in the presence of diphenyl disulfide as a devulcanizing reagent has been reported [45]. The process devulcanizes unfilled NR vulcanizates effectively. Further, a comparison of measured sol/gel components as well as dynamic mechanical properties of the devulcanized rubber products of filled and unfilled NR vulcanizates has indicated that the presence of carbon black in the vulcamizate does not disturb the devulcanization in supercritical CO_2.

1.18.4 Some Major Rubber Products

The most important application of rubbers is in the transport sector, with tires and related products consuming nearly 70% of the rubber produced. Next in importance is the application in belting for making flat conveyor and (power) transmission belts and V-belts (for power transmission), and in the hose industry for making different hoses. Rubber also has a large outlet in cellular and microcellular products. Other important and special applications of rubber are in the areas of adhesives, coated fabrics, rainwear, footwear, pipes and tubing, wire insulation, cables and sheaths, tank lining for chemical plants and oil storage, gaskets and diaphragms, rubber mats, rubber rollers, sports goods, toys and balloons, and a wide variety of molded mechanical and miscellaneous products. Formulations of a few selected rubber compounds are given in Appendix A4.

1.18.4.1 Tires

Tire technology is a very specialized area, and a tire designer is faced with the difficult task of trying to satisfy all the needs of the vehicle manufacturer, the prime factors of consideration, however, being safety and tread life.

Figure 1.75 shows the constructions of a standard bias (diagonal) ply tire and a radial ply tire. The major components of a tire are: bead, carcass, sidewall, and tread. In terms of material composition, a tire on an average contains nearly 50% of its weight in actual rubber; for oil extended rubbers (typically containing 25 parts of aromatic or cycloparaffinic oils to 75 parts of rubber), it is less. The remainder included carbon black, textile cord, and other compounding ingredients plus the beads.

The bead is constructed from a number of turns or coils of high tensile steel wire coated with copper and brass to ensure good adhesion of the rubber coating applied on it. The beads function as rigid, practically inextensible units that retain the inflated tire on the RIM.

The carcass forms the backbone of the tire. The main part of the carcass is the tire cord. The cords consist of textile threads twisted together. Rayon, nylon, and polyester cords are widely used. Steel cords are also used. While the practice of laying rubber-impregnated cords in position is still followed, the use of woven fabric is more widespread. To promote a good bond with rubber, the fiber or cord is treated with an adhesive composition such as water-soluble resorcinol-formaldehyde resin and aqueous emulsion of a copolymer of butadiene, styrene, and vinyl pyridine (70:15:15). The resincoated cord or fabric is dried and coated with a rubber compound by calendaring, whereby each cord is isolated from its neighbor. For conventional tires, the rubber-coated fabric is then cut to a predetermined width and bias angle. The bias-cut plies are joined end-to-end into a continuous length and batched into roll form, interleaved with a textile lining to prevent self-adhesion.

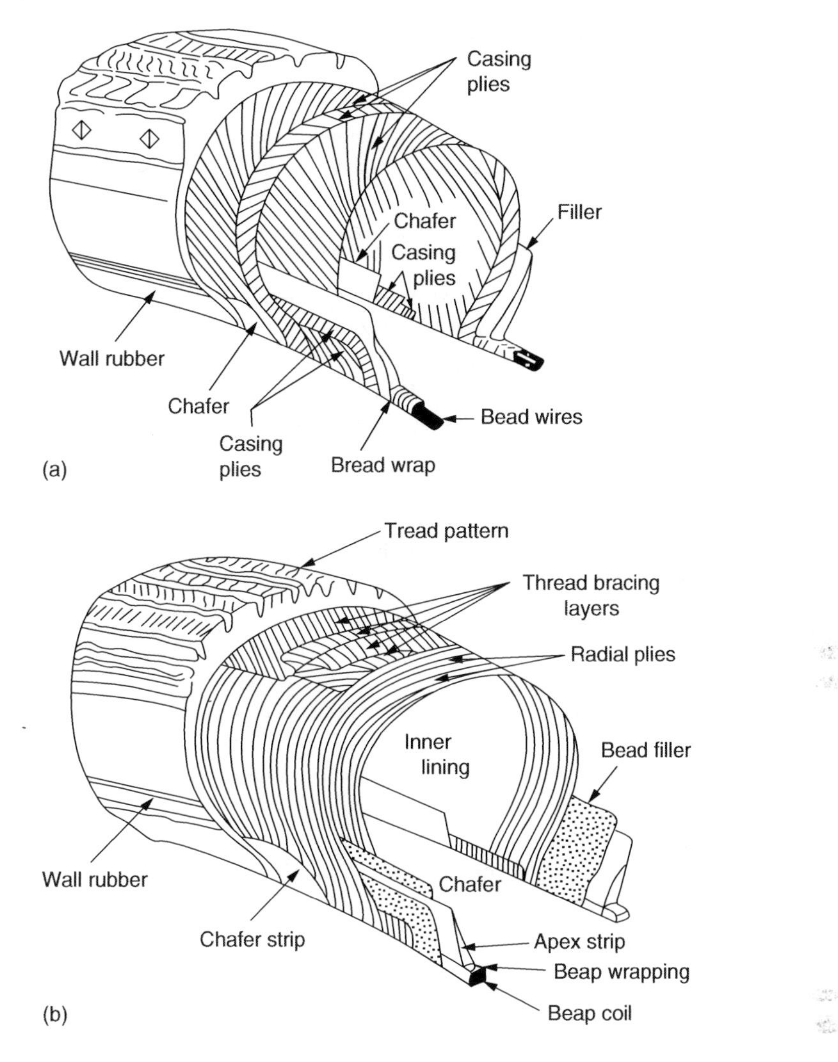

FIGURE 1.75 Diagram showing. (a) bias (diagonal) ply tire; and (b) radial ply tire.

The sidewall is a layer of extruded rubber compound that protects the carcass framework from weathering and from damage from chafing. Together with the tread and overlapping it in the buttress region, the sidewall forms the outermost layer of the tire. It is the most highly strained tire component and is susceptible to two types of failure—flex cracking and ozone cracking.

The tread is also formed by extrusion, but different rubber compounds are used for sidewall and tread (see Appendix A4). When making the side-wall and the tread in two separate extrusion lines, it is useful to take the extruded sidewall to the second extruder, which would deliver the tread to the sidewall. To simplify the tire-building operation, the sidewall and the tread may also be produced as a single unit by simultaneous extrusion of the two compounds in a single band—the tread over the sidewall—with two extruders arranged head-to-head with a common double die. The band is rapidly cooled to avoid scorching and then cut to the appropriate length. The tread receives its characteristic pattern from the mold when the subsequently built tire is vulcanized.

While building the tire, a layer of specially compounded cushion rubber may be used to keep heat development on flexing to a minimum and achieve better adhesion between the tread and the carcass.

One or more layers of fabric, known as breakers or bracing layers, may be placed below the cushion. The bracing layers raise the modulus of the tread zone and level out local blows to the tread as it contacts the road.

The tire building is carried out on a flat drum, which is rotated at a controlled speed. The plies of rubber-coated cord fabric are placed in position, one over the other, and rolled down as the drum rotates, the inner lining being placed next to the drum and the ends of the drum being flanged to suit the bead configuration of the tire. The plies of rubber-coated textile are assembled in three basic constructions—bias (diagonal), radial, and bias belted.

Bias tires have an even number of plies with cords at an angle of 30–38° from the tread center line. Passenger-car bias tires commonly have two or four plies, with six for heavy duty service. Truck tires are often built with six to twelve plies, although the larger earthmover types may contain thirty or more.

In the radial-ply tire, one or two plies are set at an angle of 90° from the center line and a *breaker* or *belt* or rubber-coated wire or textile is added under the tread. This construction gives a different tread-road interaction, resulting in a decreased rate of wear. The sidewall is thin and very flexible. The riding and steering qualities are noticeably different from those of a bias-ply tire and require different suspension systems.

The bias-belted tire, on the other hand, has much of the tread wear and traction advantage of the radial tire, but the shift from bias to bias-belted tires requires less radical change in vehicle suspension systems and in tire building machines. These features make the bias-belted tire attractive to both automobile and tire manufacturers.

When the tire building on the drum is complete, the drum is collapsed and the uncured ("green") tire is removed. The cylindrical shape of the uncured tire obtained from the building drum is transformed into a toroidal shape in the mold resulting in a circumferential stretch of the order of 60%. For shaping and curing, the uncured tire is pressed against the inner face of the heated mold by an internal bag or bladder made of a pre-cured heat-resistant rubber and inflated by a high-pressure steam or circulating hot water. Different types of molding presses are in use. In the Bag-O-Matic type of press, the uncured tire is placed over a special type of bag or bladder, and the operation of shaping, curing, and ejection of the cured tire are accomplished by an automatic sequence of machine operation.

Pneumatic tires require an air container or inner tube of rubber inside the tire. The manufacture of inner tubes is done essentially in three steps: extrusion, cutting into length, and insertion of value and vulcanization. The ends of the cut length of the tube are joined after insertion of the value prior to vulcanization.

Tubeless tires have, instead of an inner tube, an *inner liner*, which is a layer of rubber cured inside the casing to contain the air, and a chafer around the bead contoured to form an airtight seal with the RIM.

1.18.4.2 Belting and Hoses

Uses of flat conveyor and (power) transmission belts and V-belts (for power transmission) are to be found in almost all major industries. V-belts for different types cover applications ranging from fan belts for automobiles, belts for low-power drives for domestic, laboratory, and light industrial applications, to high-power belts for large industrial applications.

Textile cords or fabric and even steel cords constitute an important part of all rubber belting and hoses. The various types of cords used in the tire industry are also in use in the belting industry.

Essential steps in making conveyor belts are: (1) drying the fabric; (2) frictioning of the hot fabric with a rubber compound and topping to give additional rubber between plies and the outer ply and cover, using a three- or four-roll calender; (3) belt building; and (4) vulcanization.

The process of belt building essentially consists or cutting, laying, and folding the frictioned and coated fabric to give the desired number of plies. The construction may be straight-laid or graded-ply types (Figure 1.76) with the joints in neighboring plies being staggered to eliminate weakness and failure. Finally, the cover coat is applied by calendaring.

Vulcanization of conveyor belts may be carried out in sections using press cure or continuously by means of a Rotocure equipment. In press cure, vulcanization is done by heating in long presses, the belt being moved between successive cures by a length less than the length of the press platens. Since in this process the end or overlap zones receive an additional cure, it is desirable to minimize damage or weakness due to overcure by using flat cure mixes and allowing water cooling at each end section of the platen.

In each step of vulcanization, the section of the belt to be vulcanized is gripped and stretched hydraulically to minimize or eliminate elongation during use. The difficulties of press cure may, however, be avoided by adopting continuous vulcanization with a Rotocure equipment in which the actual curing operation is carried out between an internally steam heated cylinder and a heated steel band. Rotocure is also useful for the vulcanization of transmission belts and rubber sheeting.

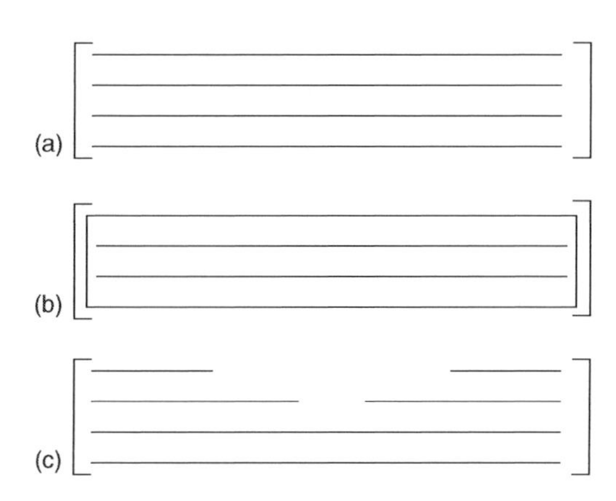

FIGURE 1.76 Construction of conveyor belts: (a) straight laid, (b) folded jacket, and (c) graded ply.

The most common feature of V-belts is their having a cross-section of the shape of a regular trapezium with the unparallel sides at an angle of 40° (Figure 1.77). The V-belts usually consist of five sections: (1) the top section known as the tension section, (2) the bottom section, called the compression section, (3) the cord section located at the neutral zone, (4) the cushion section on either side of the cord section, and (5) one or two layers of rubberized fabric, called the jacket section covering the whole assembly.

Three different rubber compounds are required for use in the above construction of a V-belt: (1) a base compound, which is the major constituent on a weight basis, (2) a soft and resilient cushion compound required for surrounding and protecting the reinforcing cord assembly, and (3) a friction compound used for rubberizing the fabric casing of the belt (see Appendix A4).

Relatively short length V-belts are built layer by layer on rotatable collapsible drum formers. The separate belts are then cut out with knives and transferred to a skiving machine that imparts the desired V-shape. The fabric jacket is then applied, and the belts are vulcanized in open steam using multi-cavity ring molds for smaller belts. Long belts are built similarly on V-groove sheave pulleys using weftless cord fabric in place of individually would cord. They are vulcanized endlessly by molding in a hydraulic press under controlled tension.

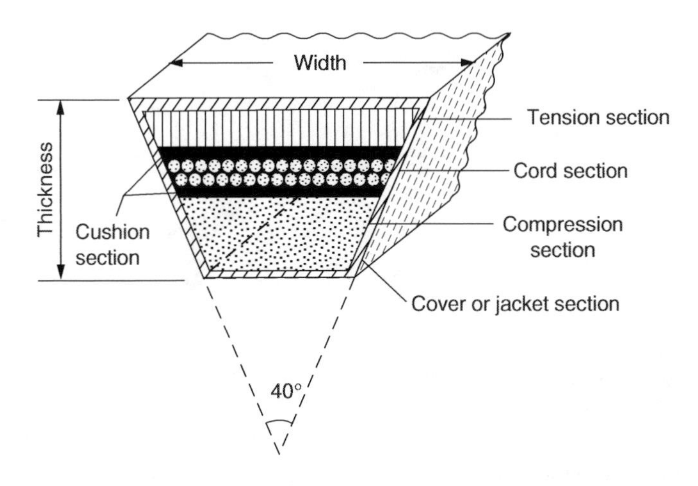

FIGURE 1.77 Cross-section showing V-belt construction.

A rubber hose has three concentric layers along the length. While the innermost part consisting of a rubber lining or tube is required to resist the action of the material that would pass through the top layer is meant to play the role of a protective layer to resist weathering, oils, chemicals, abrasion, etc. Between the inner lining and the outer cover is given a layer of reinforcement of textile yarn or steel wire applied by spiraling, knitting, braiding, or circular loom weaving. A cut woven fabric wrapped straight or on the bias may also be used to reinforce the inner lining or tube.

Essentially, the process of hose building consists of extruding the lining or tube, braiding or spiraling the textile around the cooled tube, and applying an outer cover of rubber to the reinforced hose using a cross-head extruder. Several methods are employed for vulcanization. In one process, the built hose is passed through a lead press or lead extruder to give a layer of lead cover to the hose. The hose is then wound on a drum, filled with water or air, and the ends are sealed. The whole assembly is then heated to achieve vulcanization. The water or air inside expands, and the vulcanizing hose is pressed against the lead acting as the mold. On completion of cure, the sealed ends are cut open, the lead cover is removed by slitting lengthwise in a stripping machine, and the cured hose is coiled up.

1.18.4.3 Cellular Rubber Products

Cellular rubber may be described as an assembly of a multitude of cells distributed in a rubber matrix more or less uniformly. The cells may be interconnected (open cells) as in a sponge or separate (closed cells). Foam rubber made from a liquid starting material such as latex, described earlier, is of open-cell type. Cellular products made from solid rubber are commonly called sponge (open cell structure) and expanded rubber (closed cell structure).

The technology of making cellular products from solid rubber is solely dependent on the incorporation of a blowing agent, usually a gas such as nitrogen or a chemical blowing agent, into the rubber compound. The most widely used chemical blowing agent for this application is dinitroso pentamethylene tetramine (DNPT).

The curing is carried out either freely using hot or steam or in a mold that is only partially filled with the molding compound. Synthetic rubbers, particularly SBR, are preferred as they allow precise control over level of viscosity required for obtaining consistent product quality. The sponge and expanded rubber products include carpet backing, sheets, profiles, and molding.

The development of microcellular rubber has brought a revolution in footwear technology. Microcellular rubber with an extremely fine noncommunicating cell structure and very comfortable wearing properties, is the lightest form of soling that can be produced. Density of soling as low as 0.3 g/cm^3 may be obtained with a high dose (8–10 phr) of DNPT at a curing temperature of 140–150°C. For common solings, the density normally varies between 0.5 and 0.8 g/cm^3.

High hardness and improved abrasion resistance along with low density, desired in microcellular soling, can be achieved by using SBR and high styrene resins with NR in right proportions. Higher proportions of high styrene resins give products of higher hardness and abrasion resistance and lower density. Silicious fillers such as precipitated silica and aluminum or calcium silicate also give higher hardness, abrasion resistance, and split tear strength.

Microcellular crumbs can be used in considerable quantity along with china clay and whiting to reduce the product cost. Higher proportions of stearic acid (5–10 phr) are normally used in microcellular compounds in order to bring down the decomposition temperature of DNPT type blowing agents (see Appendix A4). Post-cure oven stabilization of the microcellular sheets, typically at 100°C for 4 h, reduces the delayed shrinkage after cure to a minimum.

1.19 Miscellaneous Processing Techniques

1.19.1 Coating Processes

A coating is thin layer of material used to protect or decorate a substrate. Most often, a coating is intended to remain bonded to the surface permanently, although there are strippable coatings which are used only to afford temporary protection. An example of the latter type is the strippable hotmelt coating with ethyl cellulose as the binder [46], which is used to protect metal pieces such as drill bits or other tools and gears from corrosion and mechanical abrasion during shipping and handling.

Two of the principal methods of coating substrates with a polymer, namely extrusion coating and calendaring have already been dealt with in this chapter. Other methods of coating continuous webs include the use of dip, knife, brush, and spray. Dip coating, as applied to PVC, has already been described in previous section on plastisols.

In knife coating the coating is applied either by passing the web over a roll running partly immersed in the coating material or by running the coating material onto the face of the web while the thickness of the coating is controlled by a sharply profiled bar (or knife). This technique, also referred to as spreading, is used extensively for coating fabrics with PVC. The PVC is prepared in the form of a paste, and more than one layer is usually applied, each layer being gelated by means of heat before the next layer is added.

Lacquers are a class of coatings in which film formation results from mere evaporation of the solvents (s). The term "lacquer" usually connotes a solution of a polymer. Mixtures of solvents and diluents (extenders which may not be good solvents when used alone) are usually needed to achieve a proper balance of volatility and compatibility and a smooth coherent film on drying. Some familiar examples of lacquers are the spray cans of touch-up paint sold to the auto owner. These are mostly pigmented acrylic resins in solvents together with a very volatile solvent [usually dichlorodifluoromethane (CCl_2F_2)] which acts as a propellant. A typical lacquer formulation for coating steel surfaces contains polymer, pigment, plasticizer (nonvolatile solvent), and volatile solvents.

Latex paints or *emulsion paints* are another class of coatings which form films by loss of a liquid and deposition of a polymer layer. The paints are composed of two dispersions: (1) a resin dispersion, which is either a latex formed by emulsion polymerization or a resin in emulsion form, and (2) a dispersion of colorants, fillers, extenders, etc., obtained by milling the dry ingredients into water. The two dispersions are blended to produce an emulsion paint. Surfactants and protective colloids are added to stabilize the product.

Emulsion paints are characterized by the fact that the binder (polymer) is in a water-dispersed form, whereas in a solvent paint it is in solution form. In emulsion systems the external water phase controls the viscosity, and the molecular weight of the polymer in the internal phase does not affect it, so polymers of high molecular weight are readily utilized in these systems. This is an advantage of emulsion paints.

The minimum temperature at which the latex particles will coalesce to form a continuous layer depends mainly on the T_g. The T_g of a latex paint polymer is therefore adjusted by copolymerization or plasticization to a suitable range. The three principal polymer latexes used in emulsion paints are styrene-butadiene copolymer, poly(vinyl acetate), and acrylic resin.

Although the term "paint" has been used for latex-based systems as well as many others, traditionally it refers to one of the oldest coating systems known—that of a pigment combined with a drying oil, usually a solvent. Drying oils (e.g., linseed, tung), by virtue of their multiple unsaturation, behave like polyfunctional monomers which can polymerize ("dry") to produce film by a combination of oxidation and free-radical propagation. Oil-soluble metallic soaps are used to catalyze the oxidation process.

Combinations of resins with drying oils yield oleoresinous varnishes, whereas addition of a pigment to a varnish yields an enamel. The combination of hard, wear-resistant resin with softer, resilient, drying-oil films can be designed to give products with a wide range of durability, gloss, and hardness. Another route to obtaining a balanced combination of these properties is the *alkyd resin*, formed from *alcohols* and *acids* (and hence the "alkyd").

Alkyds are actually a type of polyester resin and are produced by direct fusion of glycerol, phthalic anhydride, and drying oil at 410°F–450°F (210°C–232°C). The process involves an esterification reaction of the alcohol and the anhydride and transesterification of the drying oil. A common mode of operation today is thus to start with the free fatty acids from the drying oil rather than with the triglycerides.

1.19.1.1 Fluidized Bed Coating

Fluidized bed coating is essentially an adaptation of dip coating and designed to be used with plastics in the form of a powder of fine particle size. It is applied for coating metallic objects with plastics. The uniqueness of the process lies in the fact that both thermoplastics and thermosetting resins can be used for the coating. Uniform coating of thicknesses from 0.005 to 0.080 in. (0.13–2.00 mm) can be built on many substrates such as aluminum, carbon steel, brass, and expanded metal. The coating is usually applied for electrical insulation and to enhance the corrosion resistance and chemical resistance of metallic parts.

Low-melting polymers are most appropriate for fluidized beds. Higher melting polymers must have a sufficiently low melt viscosity that the particles can flow and fuse together to form a continuous coating. Thermoplastic polymers in common use include nylon, PVC, acrylics, polyethylene, and polypropylene. Other possible materials are thermoplastic urethane, silicones, EVA, polystyrene, or any other low-melting, low-viscosity polymer. Thermosetting polymers are limited largely to epoxy and epoxy/polyester hybrids since other thermosets, such as phenolic and urea-formaldehyde resins, give off volatile by-products that can create voids in the coating.

The powder resin particles range in size form about 20–200 microns. Particles larger than 200 microns are difficult to suspend. Particles smaller than 20 microns may create excessive dusting and release of particles from the top of the bed.

The actual coating process is uncomplicated; however, achievement of a uniform coating requires considerable skill. The hot metal object is dipped into powdered resin fluidized by the passage of air through a porous plate. The bed is not heated; only the surface of the object to be coated is hot. As the powder contacts the hot substrate, the particles adhere, melt, and flow together to form a continuously conforming coating. The object is removed from the bed when the desired coating thickness is obtained. On cooling, the coated resin resumes its original characteristics. In the case of a thermosetting resin, additional time at elevated temperature may be required to complete the cure.

The ability of the fluidized bed to continuously conform to and coat parts having unusual shapes and sizes permits a high degree of flexibility in application. This is invaluable to the processor who wishes to coat one-of-a-kind products. In general, the coatings are smooth and glossy, with excellent adhesion to the substrate, providing the hermetic seal necessary for proper maintenance.

Fluidized bed coating is certainly the most efficient method of applying a thick coating in a single step. Thicknesses of 0.1 in.(2.5 mm) or greater are easily attained. Probably the single biggest advantage of powder coating is the nearly 100% utilization of the coating resin, without the hazard or expense of solvents.

Almost all the coatings applied by this process have a definite function, chiefly electrical insulation, but it can be used for applications that simply require a thick coating with powder. Examples of electrical applications include small motor stators and rotors, electronic components (capacitors or resistors), transformer casings, covers, laminations, and busbar. Other items coated using fluidized beds include valve bodies used in chemical industries, refinery equipment, and appliance and pump parts.

1.19.1.2 Spray Coating

Spray coating is especially useful for articles that are too large for dip coating or fluidized bed coating. The process consists of blowing out fine polymer powders through a specially designed burner nozzle, which is usually flame heated by means of acetylene or some similar gas, or it can be heated electrically. Compressed air or oxygen is used as the propelling force for blowing the polymer powder.

1.19.1.3 Electrostatic Spraying

The electrostatic spraying of polymer powders utilizes the principle that oppositely charged particles attract, a principle that has been used for many years in spraying solvent-based paints. In electrostatic spraying, polymer powder is first fluidized in a bed to separate and suspend the particles. It is then transferred through a hose by air to a specially designed spray gun. As the powder passes through the gun, direct contact with the gun and ionized air applies an electrostatic charge to the particles of powder. The contact area may be a sleeve that extends the length of the gun or merely small pins that extend into the passageway of the powder. For safety the gun is designed for high voltage but low amperage.

The part to be coated is electrically grounded, attracting the charged particles. This produces a more even coating and reduces overspray. Parts to be coated may be preheated, thereby forming the coating immediately, or they may be coated cold as the electrostatic charge will hold the particles in place until heat is applied. Once heat is applied, the particles melt and flow together, forming a continuous protective coating.

Coatings 50–75 microns thick can be applied electrostatically to cold objects and coatings up to 250 microns thick to hot objects. The polymers used in spraying of powders are the same as those used in fluidized beds. The key characteristic for any polymer, thermoplastic or thermoset, applied as a powder is low melt viscosity, which enables polymer particles to flow together and form a continuous coating.

The chief characteristic of electrostatically applied powder polymer is the ability to produce a thin coating. It is the preferred method of producing a coating of 0.001–0.002 in. (0.025–0.050 mm) thickness. It is a continuous process and suited to automated assembly-line production. With a well designed recovery system and fully enclosed spray booth, the process permits full utilization of the powder. The particle size of sprayed powders is smaller than that of powders used in fluidized bed coating. The average particles size is 30–60 microns.

Much of what is termed decorative powder coating is applied by electrostatic spray. Appliances, laboratory instruments, transformer housings, engine parts, and chain-link fences are among the many types of products so coated for decorative purposes.

Coatings that provide electrical insulation are also applied by electrostatic spray. Objects coated include electrical motor armatures and stators, electrical switchgear boxes, and magnet wire. Corrosion-resistant coatings are designed to prevent the corrosion of the underlying substrates. These include pipe, fencing, concrete reinforcing bars, valves, conduit, and pumps. Polymers used for corrosion protection are usually thermosetting, particularly epoxies because of their superior adhesion characteristics.

1.19.2 Powder Molding of Thermoplastics

1.19.2.1 Static (Sinter) Molding

The process is often used with polyethylene and is limited to making open-ended containers. The mold which represents the exterior shape of the product is filled with powder. The filled mold is heated in an oven, causing the powder to melt and thus creating a wall of plastics on the inner surface of the mold. After a specific time to build the required wall thickness, the excess powder is dumped from the mold and the mold is returned to the oven to smooth the inner wall. The mold is then cooled, and the product is removed. The product is strain free, unlike pressure-molded products. In polyethylene this is especially significant if the product is used to contain oxidizing acids.

1.19.2.2 Rotational Molding

Rotational molding (popularly known as rotomolding) is best suited for large, hollow products, requiring complicated curves, uniform wall thickness, a good finish, and stress-free strength. It has been used for a variety of products such as car and truck body components (including an entire car body), industrial containers, telephone booths, portable outhouses, garbage cans, ice buckets, appliance housing, light globes, toys, and boat hulls. The process is applicable to most thermoplastics and has also been adapted for possible use with thermosets.

Essentially four steps are involved in rotational molding: loading, melting and shaping, cooling, and unloading. In the loading stage a predetermined weight of powdered plastic is charged into a hollow mold. The mold halves are closed, and the loaded mold is moved into a hot oven where it is caused to simultaneously rotate in two perpendicular planes. A 4:1 ratio of rotation speeds on minor and major

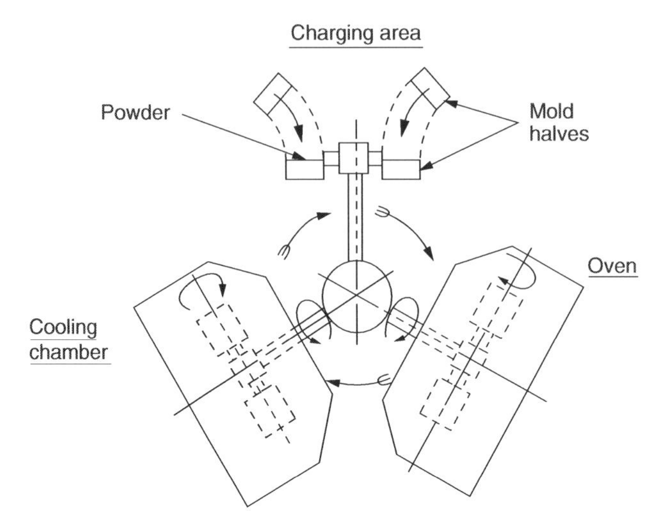

FIGURE 1.78 Basics of continuous-type three-arm rotational molding machine.

axes is generally used for symmetrically shaped objects, but wide variability of ratios is necessary for objects having complicated configurations.

On most units the heating is done by air or by a liquid of high specific heat. The temperature in the oven may range from 500°F–900°F (260°C–482°C), depending on the material and the product. As the molds continue to rotate in the oven, the polymer melts and forms a homogenous layer of molten plastic, distributed evenly on the mold cavity walls through gravitational force (centrifugal force is not a factor). The molds are then moved, while still rotating, from the oven into the cooling chamber, where the molds and contents are cooled by forced cold air, water fog, or water spray. Finally, the molds are opened, and the parts removed. (Rotational molding of plastisol, described earlier, is similar to that described here. However, in this case the plastic is charged in the form of liquid dispersion, which gels in the cavity of the rotating mold during the heating cycle in the oven.)

The most common rotational molding machine in use today is the carousel-type machine. There are three-arm machines consisting of three cantilevered arms or mold spindles extending from a rotating hub in the center of the unit. In operation, individual arms are simultaneously involved in different phases (loading, heating, and cooling) in three stations so that no arms are idle at any time (Figure 1.78).

Modern rotational-molding machines enable large parts to be molded (e.g., 500-lb car bodies and 500-gal industrial containers). For producing small parts an arm of a carousel-type machine may hold as many as 96 cavities.

1.19.2.3 Centrifugal Casting

Centrifugal casting is generally used for making large tube forms. It consists of rotating a heated tube mold which is charged uniformly with powdered thermoplastic along its length. When a tubular molten layer of the desired thickness builds up on the mold surface, the heat source is removed and the mold is cooled. The mold, however, continues to rotate during cooling, thus maintaining uniform wall thickness of the tube. Upon completion of the cooling cycle, the plastic tube, which has shrunk away from the mold surface, is removed, and the process is repeated. Usual tube sizes molded by the process range from 6–30 in. in diameter and up to 96 in. in length.

1.19.3 Adhesive Bonding of Plastics

Adhesives are widely used for joining and assembling of plastics by virtue of low cost and adaptability to high-speed production. They can be subdivided into solvent or dope cements, which are suitable for most thermoplastics (not thermosets), and monomeric or polymerizable cements which can be used for most thermoplastics and thermosets.

Solvent cements and dope cements function by attacking the surfaces of the adherends so that they soften and, on evaporation of the solvent, will join together. The dope cements, or bodies cements, differ from the straight solvents in that they contain, in solution, quantity of the same plastic which is being bonded. On drying, these cements leave a film of plastic that contributes to the bond between the surfaces to be joined.

Monomeric or polymerizable cements consist of a reactive monomer, identical with or compatible with the plastic to be bonded, together with a suitable catalyst system and accelerator. The mixture polymerizes either at room temperature or at a temperature below the softening point of the thermoplastic being joined. In order to hasten the bonding and to reduce shrinkage, some amount of the solid plastic may also be initially dissolved in the monomer. Adhesives of this type may be of an entirely different chemical type than the plastic being joined. A typical example is the liquid mixture of epoxy resins and hardener which by virtue of the chemical reactivity and hydrogen bonding available from the epoxy adhesive provide excellent adhesion to many materials.

In joining with solvents or adhesives, it is very important that the surfaces of the joint be clean and well matched since poor contact of mating surfaces can cause many troubles. The problem of getting proper contact is aggravated by shrinkage, warpage, flash, marks from ejector pins, and nonflat surfaces.

1.19.3.1 Solvent Cementing

Solvent cementing or solvent welding basically involves softening the bonding area with a solvent or a solvent containing small quantities of the parent plastic, referred to as dope or bodied cement, generally containing less than 15% resin. The solvent or cement must be of such composition that it will dry completely without blushing. Light pressures must be applied to the cemented joint until it has hardened to the extent that there is no movement when released. Structural bonds of up to 100% of the strength of the parent material are possible with this type of bonding.

Table 1.8 gives a list of typical solvents selected as useful for cementing of plastics. A key to selection of solvents in this table is how fast they evaporate: a fast-evaporating solvent may not last long enough for some assemblies, while evaporation that is too slow could hold up production. It may be noted that solvent bonding of dissimilar materials is possible where the materials can be bonded with the same solvents (see Table 1.8).

The Solvent is usually applied by the soak method in which pieces are immersed in the solvent, softened, removed, quickly brought together, and held under light pressure for some time. Areas adjoining the joint area should be masked to prevent them from being etched. For some applications where the surfaces to be cemented fit very closely, it is possible to introduce the cementing solvent by brush, eyedropper, or hypodermic needle into the edges of the joint. The solvent is allowed to spread to the rest of the joint by capillary action.

TABLE 1.8 Typical Solvents for Solvent Cementing of Plastics

Plastics	Solvent
ABS	Methylene chloride, tetrahydrofuran, methyl ethyl ketone, methyl isobutyl ketone
Acetate	Methylene chloride, chloroform, acetone, ethyl acetate, methyl ethyl ketone
Acrylic	Methylene chloride, ethylene dichloride
Cellulosics	Acetone, methyl ethyl ketone
Nylon	Aqueous phenol, solutions of resorcinol in alcohol, solutions of calcium chloride in alcohol
PPO	Methylene chloride, chloroform, ethylene dichloride, trichloroethylene
PVC	Cyclohexane, tetrahydrofuran, dichlorobenzene
Polycarbonate	Methylene chloride, ethylene dichloride
Polystyrene	Methylene chloride, ethylene dichloride, trichloroethylene, methyl ethyl ketone, xylene
Polysulfone	Methylene chloride

1.19.3.2 Adhesive Bonding

Adhesive bonding is a process in which the adhesive acts as an agent to hold two substrates together (as opposed to solvent cementing where the parent materials actually become an integral part of the bond) and the adhesion is chemical.

Table 1.9 gives a list of various adhesives and typical applications in bonding of plastics. This table is not complete, but it does give a general idea of what types of adhesives are used and where. It should be remembered, however, that thousands and thousands of variations of standard adhesives are available off the shelf. The computer may shape up as an excellent selection aid for adhesives. Selection is made according to the combination of properties desired, tack time, strength, method of application, and economics (performance/cost ratio).

The form that the adhesive takes (liquids, mastics, hot melts, etc.) can have a bearing on how and where they are used. The anaerobics, which can give some very high bond strength and are usable with all materials except polyethylene and fluorocarbons, are dispensed by the drop. A thin application of the anaerobics is said to give better bond strength than a thick application.

The more viscous, mastic-type cements include some of the epoxies, urethanes and silicones. Epoxies adhere well to both thermosets and thermoplastics. But epoxies are not recommended for most polyolefin bonding. Urethane adhesives have made inroads into flexible packaging, the shoe industry, and vinyl bonding. Polyester-based polyurethanes are often preferred over polyether systems because of their higher cohesive and adhesive properties. Silicones are especially recommended where both bonding and sealing are desired.

Hot melts—100% solids adhesives that are heated to produce a workable material—are based on polyethylene, saturated polyester or polyamide in chunk, granule, pellet, rope, or slug form. Saturated polyesters are the primary hot melts for plastics; the polyethylenes are largely used in packaging; polyamides are used most widely in the shoe industry. Application speeds of hot melts are high and it pays to consider hot melts if production requirements are correspondingly high.

Pressure sensitives are contact-bond adhesives. Usually rubber based, they provide a low-strength, permanently tacky bond. They have a number of consumer applications (e.g., cellophane tape), but they are also used in industrial applications where a permanent bond is not desirable or where a strong bond may not be necessary. The adhesive itself is applied rapidly by spray. Assembly is merely a matter of pressing the parts together.

Film adhesives require an outside means such as heat, water, or solvent to reactivate them to a tacky state. Among the film types are some hot melts, epoxies, phenolics, elastomers, and polyamides. Film adhesives can be die cut into complicated shapes to ensure precision bonding of unusual shapes. Applications for this type of adhesive include bonding plastic bezels onto automobiles, attaching trim to both interiors and exteriors, and attaching nameplates on luggage.

While most plastics bond without trouble once the proper adhesive has been selected, a few, notably polyolefins, fluorocarbons and acetals, require special treatment prior to bonding. Untreated polyolefins adhere to very few substrates (a reason that polyethylene is such a popular material for packaging adhesives). Their treating methods include corona discharge, flame treating (especially for large, irregular-shaped articles), and surface oxidation by dipping the articles in a solution of potassium dichromate and sulfuric acid. Fluorocarbons can be prepared for bonding by cleaning with a solvent such as acetone and then treating with a special etching solution. For acetals, one pretreatment method involves immersing articles in a special solution composed mainly of perchloroethylene, drying at 120°C (250°F), rinsing, and then air drying.

1.19.3.3 Joining of Specific Plastics

The same basic handling techniques apply to almost all thermoplastic materials. In the following section, however, a few thermoplastics will be treated separately, with mention of the specific cements most suitable for each. The section on acrylics should be read in connection with the cementing of any other thermoplastics.

TABLE 1.9 Typical Adhesives for Bonding Plastics

	ABS	Acetal	Acrylic	Cellulosics	Fluorocarbons	Nylon	PVC	PC	PE	PP	PS	MF	PF	Polyesters	PU
Metals	10	3,10	2,3	2,3	9,10	2,10	2,3,7,10,13,14	10	2,12	1	12	3,14	2	4	3,4
Paper	—	3,10	—	—	9,10	3	—	13	—	1	4,12,13	—	—	—	4,13
Wood	10	10	2,3	3	10	2,3	3,10,13	10,13	2	1	12,13	2,3	2	2	13
Rubber	10	3	1–4	1–4	10	2	3,4,7	4,13	2	1	5	2,3,14	2,3	1,4	4,13
Ceramics	—	10	2,3	3	10	3,10	3,4	10,13	2	1	—	2	2	2	3
ABS	10,14	2,4,6,10	2,4,6,10	—	—	8,10	10	4,10	10	—	—	—	—	—	—
Acetal	2,4,6,10	3,10	10	—	10	10	10	—	—	—	—	—	10	—	—
Acrylic	2,4,6,10	2,4,6,10	15	—	10	2,3,4,9	3,4,6,10	—	—	—	—	—	4,8,10	10	4,10
Cellulosic	—	—	—	3,4,13	10	2,3,7,9	10	—	—	—	—	—	10	—	—
Fluorocarbons	—	10	—	—	9,10	—	10	—	10	—	10	—	—	—	10
Nylon	8,10	8,10	2,3,7,9	2,3,7,9	—	2,9,10,13	—	—	—	—	—	8,10	8,10,14	8,10	—
PVC	10	10	3,4,6,10	—	10	—	3,4,6,13	—	—	—	—	—	4,7	10	—
PC	10	—	—	—	—	—	3,4,6,13	15	—	—	—	—	—	—	—
PE	10	—	—	—	10	—	—	—	10,12	—	—	—	—	—	—
PP	—	—	—	—	—	—	—	—	—	10,12	—	—	—	—	—
PS	—	—	—	—	10	—	—	—	—	—	4,5,10,12,13	—	4,10,14	—	—
MF	—	—	—	—	—	—	—	—	—	—	—	2,3,10,12,13	—	—	—
PF	10,14	14	4,8,10	10	—	4,7	—	—	—	4,10	3,8,10,11	2–4,10,12,14	—	—	—
Polyester	—	—	10	10	—	—	10	—	—	—	—	—	—	3,10,12,13	—
PU	—	—	4,10	—	10	—	—	—	—	—	—	—	—	—	3,4,10,13

Elastomeric: 1, Natural rubber. 2, Neoprene. 3, Nitrile. 4, Urethane. 5, Styrene-butadiene. Thermoplastic: 6, Poly(vinyl acetate). 7, Polyamide. Thermosetting: 8, Phenol-formaldehyde. 9, Resorcinol, Phenol-resorcinol/formaldehyde. 10, Epoxy. 11, urea-formaldehyde. Resin: 12, Phenolic-poly(vinyl butyral). 13, Polyeser. Other: 14, Cyanoacrylate. 15, Solvent. *Source:* Adapted from O'Rinda Trauernicht, J. 1970. *Plastics Technology*, Reinhold Publishing, New York.

1.19.3.3.1 Cast Acrylic Sheeting

Articles of considerable size and complexity can be fabricated from methyl methacrylate plastics by joining sections together by solvent welding. The technique described here applies to cast sheeting. With articles made from methyl methacrylate molding powders or extruded rod, tubing or other shapes, joining is generally not as satisfactory as with cast sheeting.

With care and practice, the transparency of acrylic resin can be retained in joints with the formation of a complete union of the two surfaces brought into contact. Usually one of the two surfaces to be joined is soaked in the cementing solvent until a soft, swollen layer (cushion) has been formed upon it. This soft surface is then pressed against the surface to be attached and held in contact with it so that the excess solvent contained in the soaked area softens it also.

For some purposes, it may be desirable to dissolve clean savings of methyl methacrylate resin in the solvent in order to raise its viscosity so that it can be handled like glue.

The most universally applicable type of solvent cement is the polymerizable type, comprising a mixture of solvent and catalyzed monomer. These are mobile liquids, volatile, rapid in action, and capable of yielding strong sound bonds. An example of these is a 40–60 mixture of methyl methacrylate monomer and methylene chloride. Before using this cement 1.2 grams of benzoyl peroxide per pint of solvent, should be added. Heat treatment or annealing of joints made with solvent cements is highly desirable because it greatly increases the strength of the joint.

1.19.3.3.2 Cellulosics

The cements used with cellulosic plastics are of two types: (1) solvent type, consisting only of a solvent or a mixture of solvents; (2) dope type, consisting of a solution of the cellulosic plastic in a solvent or mixture of solvents.

Acetone and mixture of acetone and methyl "cellosolve" are commonly used as solvent cements for cellulose acetate. Acetone is a strong solvent for the plastic, but evaporates rapidly. The addition of methyl "cellosolve" retards the evaporation, prevents blushing, and permits more time for handling the parts after application of the cement.

A cement of the dope type leaves upon drying a film of plastic that forms the bond between the surfaces to be joined. These cements are generally used when an imperfect fit of the parts requires filling. A typical composition of the dope-type cement for cellulose acetate is:

	Parts by Weight
Cellulose acetate	130
Acetone	400
Methyl "cellosolve"	150
Methyl "cellosolve" acetate	50

Other cellulosics, cellulose acetate butyrate and propionate are cemented in accordance with the technique described for cellulose acetate. In the case of dope cements, the plastic to be dissolved in solvents is cellulose propionate. Similarly for ethyl cellulose plastic, the strongest bonds are made by solvents or by solvents bodied with ethyl cellulose plastic.

1.19.3.3.3 Nylon

The recommended cements for nylon-to-nylon bonding are generally solvents, such as aqueous phenol, solutions of resorcinol in alcohol, and solutions of calcium chloride in alcohol, sometimes "bodied" by the inclusion of nylon in small percentages.

Aqueous phenol containing 10–15% water is the most generally used cement for bonding nylon to itself. The bond achieved by use of this cement is water resistant, flexible, and has strength approaching that of the nylon.

Calcium–chloride–ethanol solution bodied with nylon is recommended for nylon-to-nylon joints where there is possibility of contact with foods or where phenol or resorcinol would be otherwise objectionable.

For bonding nylon to metal and other materials, various commercial adhesives, especially those based on phenol-formaldehyde and epoxy resins, are sometimes used. Epoxy adhesives (in two-part systems), for example, have been used to produce satisfactory joints between nylon and metal, wood, glass and leather.

1.19.3.3.4 Polycarbonate

Solvent cementing of parts of polycarbonate may be effected by the use of a variety of solvents or light solutions of polycarbonate in solvents. Methylene chloride, a 1–5% solution of polycarbonate in methylene chloride, and a mixture of methylene chloride and ethylene dichloride (with a maximum of 40% ethylene dichloride) are commonly recommended.

Solvent should be applied to only one of the bonding surfaces while the other half remains dry and ready in the clamping fixture. As soon as the two parts have been put together, pressure should be applied immediately. Pressure between 200 and 600 psi is suggested for best results. Holding time in the pressure fixture is approximately 1–5 min, depending on the size of the bonding area.

For bonding molded parts of polycarbonate to other plastics, glass, wood, aluminum, brass, steel, and other materials, a wide variety of adhesives can be used. Generally, the best results are obtained with solventless materials, such as epoxies and urethanes.

1.19.3.3.5 Polyethylene

The good solvent resistance of polyethylene and other olefins precludes the use of solvent-type cements. Several commercial rubber-type adhesives produce moderate adhesion with polyethylene that has been surface treated. One technique for surface treatment is to dip polyethylene in a chromic acid bath (made up of concentrated sulfuric acid 150 parts by weight, water 12 parts, and potassium dichromate 7.5 parts) for about 30 sec at 70°C. The parts are rinsed with water after this treatment. Still another effective surface treatment for producing cementable surfaces on polyethylene is electrical discharge. The open oxidizing flame method is also used extensively for this purpose.

1.19.3.3.6 Polystyrene

Complex assemblies of polystyrene, usually molded in section, may be joined by means of solvents and adhesives. Polystyrene is soluble in a wide variety of solvents. According to their relative volatilities they may be divided into three groups—fast drying, medium drying, and slow drying.

Methylene chloride, ethylene dichloride, and trichloroethylene are some of the fast-drying solvents that produce strong joints. However, they are unsatisfactory for transparent articles of polystyrene because they cause rapid crazing. "Medium-drying" solvents such as toluene, perchloroethylene, and ethyl benzene that have higher boiling temperature are less apt to cause crazing.

High-boiling or "slow-drying" solvents such as amylbenzol and 2-ethylnaphthalene often require excessive time for development of sufficient bond strength, but they will not cause crazing to appear so quickly. Up to 65% of a fast- or medium-drying solvent may be added to a slow-drying solvent to speed up the development of initial tack without greatly reducing the time before crazing appears.

A bodied, or more viscous, solvent may be required by certain joint designs and for producing airtight or watertight seals. These are made by dissolving usually 5%–15% of polystyrene by weight in a solvent. Solvent-based contact cements provide the strongest bond between polystyrene and wood. These adhesives all have a neoprene (polychloroprene) base and a ketonic-aromatic solvent system.

1.19.3.3.7 Poly(Vinyl Chloride) and Copolymers

On account of the relative insolubility of PVC and the markedly increased effect of solvents with the increasing content of vinyl acetate in the copolymer resins, there exists among the vinyl chloride-acetate copolymer system a great diversity of composition and of ability to be cemented by solvents.

The copolymer resins are most rapidly dissolved by the ketone solvents, such as acetone, methyl ethyl ketone, methyl isobutyl ketone, and cyclohexanone. Propylene oxide also is a very useful solvent in hastening solution of copolymer resins, especially those of high molecular weight, and of straight PVC. This solvent penetrates the resins very rapidly and, in amounts up to about 20–25%, improves the "bite" into the resin. The chlorinated hydrocarbons also are excellent solvents for the vinyl chloride-vinyl acetate copolymer resins, and are suitable for use in cements.

1.19.3.3.8 Thermosetting Plastics

Adhesive bonding is, for various reasons, the logical method of fastening or joining cross-linked and reinforced thermoset plastics to themselves, or to other materials. Since most thermoset plastics are quite resistant to solvents and heat, heat-curing solvent-dispersed adhesives may be used. Such adhesives consist of reactive or thermosetting resins (e.g., phenolics, epoxies, urea-formaldehydes, alkyds, and combinations of these), together with compatible film-formers such as elastomers or vinyl-aldehyde condensation resins. Isocyanates are frequently added as modifiers to improve specific adhesion to surfaces that are difficult to bond. These adhesives may be applied not only in solvent-dispersed form, but also in the form of film, either unsupported, or supported on fabric, glass-mat, and so forth.

A great many of outstanding adhesive formulations are based on epoxy resins. A broad spectrum of adhesive formulations with a wide range of available properties have resulted from the use of polymeric hardeners such as polyamides and polyamines, phenolics, isocyanates, alkyds, and combinations of amines with polysulfide elastomers, and the "alloying" of the epoxy with compatible polymeric film-formers, such as poly(vinyl acetate) and certain elastomers.

In cemented assemblies of thermoset plastics and metals, where structural strength is generally desired, the adhesive must be more rigid than those used for bonding plastic to plastic, i.e., one with modulus, strength, and coefficient of thermal expansion between those of the plastic and the metal. In many cases, such adhesives are stronger than the plastic itself.

1.19.4 Welding

Often it is necessary to join two or more components of plastics to produce a particular setup or to repair a broken part. For some thermoplastics solvent welding is applicable. The process uses solvents which dissolve the plastic to provide molecular interlocking and then evaporate. Normally it requires close-fitting joints. The more common method of joining plastics, however, is to use heat, with or without pressure. Various heat welding processes are available. Those processes in common commercial use are described here.

1.19.4.1 Hot-Gas Welding

Hot-gas welding, which bears a superficial resemblance to welding of metals with an oxyacetylene flame, is particularly useful for joining thermoplastic sheets in the fabrication of chemical plant items, such as tanks and ducting. The sheets to be joined are cleaned, beveled, and placed side by side so that the two beveled edges form a V-shaped channel. The tip of a filler rod (of the same plastic) is placed in the channel, and both it and the adjacent area of the sheets are heated with a hot-gas stream (200–400°C) directed from an electrically heated hot-gas nozzle (Figure 1.79a), which melts the plastics. The plastics then fuse and unite the two sheets. The hot gas may be air in PVC welding, but for polyethylene an inert gas such as nitrogen must be used to prevent oxidation of the plastics during welding.

1.19.4.2 Fusion Welding

Fusion or hot-tool welding is accomplished with an electrically heated hot plate or a heated tool (usually of metal), which is used to bring the two plastic surfaces to be joined to the required temperature. The polyfusion process for joining plastic pipes by means of injection-molded couplings is an example of this type of welding. The tool for this process is so shaped that one side of it fits *over* the pipe while the other side fits into the coupling. The tool is heated and used to soften the outside wall of the pipe and the inside wall of the coupling. The pipe and coupling are firmly pressed together and held until the joint cools to achieve the maximum strength of the weld. The tool is chrome plated to prevent the plastic sticking to its surfaces.

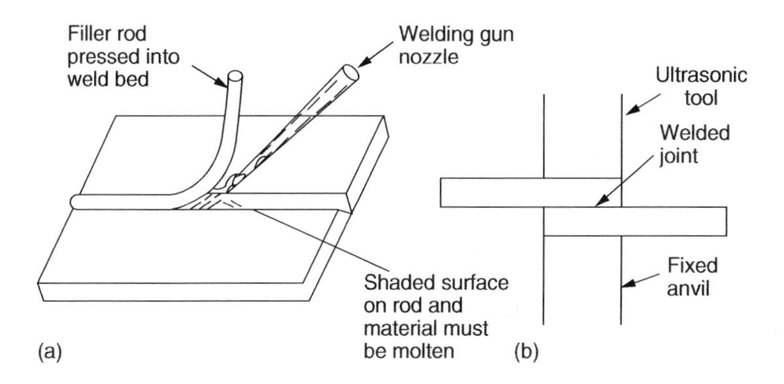

FIGURE 1.79 Welding of plastics: (a) hot-gas welding; (b) ultrasonic contact welding.

1.19.4.3 Friction Welding

In friction or spin-welding of thermoplastics, one of the two pieces to be jointed is fixed in the chuck of a modified lathe and rotated at high speed while the other piece is held against it until frictional heat causes the polymer to flow. The chuck is stopped, and the two pieces are allowed to cool under pressure. The process is limited to objects having a circular configuration. Typical examples are dual-colored knobs, molded hemispheres, and injection-molded bottle halves.

1.19.4.4 High-Frequency Welding

Dielectric or high-frequency welding can be used for joining those thermoplastics which have high dielectric-loss characteristics, including cellulose acetate, ABS, and PVC. Obviously, polyethylene, polypropylene, and polystyrene cannot be welded by this method. The device used for high-frequency welding is essentially a radio transmitter operated at frequencies between 27 and 40 MHz. The energy obtained from the transmitter is directed to electrodes of the welding apparatus. The high-frequency field causes the molecules in the plastic to vibrate and rub against each other very fast, which creates frictional heat sufficient to melt the interfaces and produce a weld.

1.19.4.5 Ultrasonic Welding

In ultrasonic welding the molecules of the plastic to be welded are sufficiently disturbed by the application of ultrahigh-frequency mechanical energy to create frictional heat, thereby causing the plastics to melt and join quickly and firmly.

The machinery for ultrasonic welding consists of an electronic device which generates electrical energy at 20/50 kHz/sec and a transducer (either magnetostrictive or piezoelectric) to convert the electrical energy to mechanical energy. In the contact-welding method (Figure 1.79b) the ultrasonic force from the transducer is transmitted to the objects (to be welded) through a tool or "horn," generally made of titanium. The amplitude of the motion of the horn is from 0.0005 to 0.005 in. (0.013–0.13 mm) depending on the design. The method is generally used for welding thin or less rigid thermoplastics, such as films, or sheets of polyethylene, plasticized PVC, and others having low stiffness.

1.20 Decoration of Plastics

Commercial techniques for decorating plastics are almost as varied as plastics themselves. Depending on end-use applications or market demands, virtually any desired effect or combination of effects, shading of tone, and degree of brightness can be imparted to flexible or rigid plastics products.

The primary decorating technique is raw-materials coloring achieved at the compounding stage. Although most thermoplastics are produced in natural white or colorless transparent form, color is usually added by directly blending colorants into the base resin prior to the processing stage. These colorants (or color concentrates) are available in a wide range of stock shades with precise tinctorial values.

Colors can also be matched to exact customer specifications and these specifications kept in computer memory to ensure batch-to-batch or order-to-order consistency. Color blending can also be utilitarian, as in color-coded wire- and cable-sheathing.

Besides basic raw-materials coloring, mentioned above, designers have a large palette of decorating media at their disposal. Plastics can be decorated in various ways, which include painting processes, direct printing, transfer decoration, in-mold decoration, embossing, vacuum metallizing, sputtering, and electroplating. Most of these processes require bonding other media, such as inks, enamels, and other materials to the plastics to be decorated.

Some plastics, notably polyolefins and acetals, are, however, highly resistant to bonding and need separate treatment to activate the surface. Commonly used treatment processes are flame treatment, electronic treatments such as corona discharge and plasma discharge, and chemical treatment.

In flame treatment, plastic objects such as bottles and film are passed through an oxidizing gas flame. Momentary contact with the film causes oxidation of the surface, which makes it receptive to material used in decorating the product.

In the corona discharge process the plastic film to be treated is allowed to pass over an insulated metal drum beneath conductors charged with a high voltage. When the electron discharge ("corona") between the charged conductors and the drum strikes the intervening film surface, oxidation occurs and makes the surface receptive to coatings. Molded products are also treated in a similar manner, often by fully automatic machinery.

In the plasma process [48], air at low pressure is passed through an electric discharge, where it is partially dissociated into the plasma state and then expanded into a closed vacuum chamber containing the plastic object to be treated. The plasma reacting with the surfaces of the plastic alters their physicochemical characteristics in a manner that affords excellent adhesion to surface coatings. The process can be used for batch processing of plastics products, including films which may be unreeled in the vacuum chamber for treatment.

Acetal resin products are surface treated by a chemical process consisting of subjecting the product to a short acid dip that results in an etched surface receptive to paint.

1.20.1 Painting

Virtually all plastics, both thermoplastic and thermosetting, can be pained, with or without priming or other preliminary preparation procedures. The process, however, requires special consideration of the resin-solvent system to achieve adhesion, adequate covering, and chemical resistance. Painting operations have the advantage of being as simple or as sophisticated as the application may dictate.

Plastics parts or materials can be coated manually by brushing, dipping, hand-spray painting, flow coating or roller coating; they can be automatically spray-painted with rotating or reciprocating spray guns, and electro-statically painted using a conductive precoating procedure.

Painting operations have the advantage of offering almost unlimited color options as well as great variety of surface finishes and final surface properties to meet such needs as gloss, UV resistance, abrasion resistance, and chemical resistance.

1.20.2 Printing

The primary printing presses used in plastics are gravure printing, flexography, silk-screen printing, and pad printing.

1.20.2.1 Gravure Printing

Gravure printing is a process that requires the use of an engraved metal cylinder or roller. Rotogravure is thus an appropriate title for this printing process. The engraving or etching process on the surfaces of the metal cylinder results in recessed areas that pick up ink or liquid coatings from a reservoir. With proper formulation of printing ink, the gravure process can be applied to a great variety of plastic substrates. Virtually all thermoplastic film or sheet applications are printable by this process.

A good example of the capabilities of the gravure process is the printing of woodgrain patterns on carrier foil for use in hot-stamping applications (described later). Woodgrain patterns may require the application of several coatings to achieve the proper effect. Several engraved cylinders can be used in sequence for continuous printing.

1.20.2.2 Flexography

Flexography uses a flexible printing plate, typically a metal-silicone rubber-bonded combination with the rubber surface processed to leave the printing surface raised over the back-ground area. The raised and recessed areas on the surface can be fabricated through photographic etching and/or engraving. After transferring the ink from a reservoir through a roller-doctor blade system onto the curved flexible plate, the ink is transferred off the raised portions to the material to be printed.

The process is suitable for a variety of applications, ranging from simple label film to decoration on molded parts such as plaques, medallions or wall tile. However, the flexible printing plates used in flexography do not permit the very fine detail that can be achieved on metal surfaces such as used in gravure printing. There are also limitations to the size and shape of articles that can be printed.

1.20.2.3 Screen Process Printing

The process derived its name from the use of silk cloth or silk screen in the transfer of printing ink to articles to be printed. Integral to the process is the use of a suitable open-weave cloth or screen (silk is still commonly but not exclusively used) stretched over a framework. Screens made of nylon or other synthetic material are often employed, as also stainless steel or other metallic screens. The stretched screen is selectively coated through the use of a stencil corresponding to an art copy of the image to be printed (see "Silkscreen Printing" in Chapter 2 of *Industrial Polymers, Specialty Polymers, and Their Applications*); this coated (closed) area resists the passage of printing ink, which can only penetrate through the uncoated (open) areas of the screen. There are various ways to prepare the screen for printing, other than stenciling.

Polyolefins such as polyethylene and polypropylene must be surface treated before being printed. The most effective way is in an integrated machine where surface treatment takes place right before printing. A time lapse will mean that the treatment will lose some of its effect. Three methods are used; flame treatment, corona discharge, and chemical treatment. Flame treatment is considered the most practical and most widely used.

1.20.2.4 Pad Printing

Pad printing uses printing principles and techniques from letterpress and flexography. The uniqueness of the process has to do with the use of a smooth silicone pad that picks up ink impression from an engraved or etched plate and transfers it to the product to be decorated. The engraved plate, known as a cliche, is produced in a manner similar to that of printing plates for offset or gravure roller printing.

The silicone pickup pad can be designed to meet almost any shape and configuration of the product part. This ability has prompted tremendous growth in pad printing. An additional capability of the process is that it can print several colors and impressions within one cycle of operation. Coatings can be layered when wet to accomplish multicolor designs with very accurate registration and impression quality.

1.20.3 Hot Stamping

Hot stamping is one of the original methods of decorating plastics materials. Though familiarly known as hot stamping, the terminology "coated foil transferring" might be more appropriate since in this process the printed coating on a carrier film is transferred onto a plastic surface. Secure adhesion is accomplished with the use of heat, pressure, and time under controlled conditions.

The key to the process is the use of a carrier film (usually a polycarbonate, polyester, or cellophane) upon which various coatings provide the desired decorative effect. The coated foil is placed over the plastic to be decorated, and a heated die forces the foil onto the plastic. The proper control of heat, pressure, and time transfers the coating off the carrier foil onto the plastic.

The hot-stamping process is a versatile tool for plastics decoration. A wide variety of coatings can be deposited on the carrier film which allows the process to be used on almost any thermoplastic material and many thermosets. Metallic effects can be imparted by depositing microthin coatings of gold or silver or chrome; multiple coatings can be applied to the carrier film to achieve such effects as woodgraining, marbleizing, or multicolored designs. Three-dimensional decorative effects can also be achieved by embossing the surface of the carrier medium coating.

1.20.4 In-Mold Decorating

As the name implies, in-mold decorating is a process in which a predecorated overlay (film), or decal, is placed in the mold, where the decorated element is fused to the molded part during the heating/cooling cycles of the molding operation. Since the decorated coating is bonded between the plastic part and the film (which will be the exterior surface), thus forming an integral part of the product, it produces one of the most durable and permanent decorations. High-quality melamine dinnerware is decorated by this method, and so are a host of other household and hardware plastics goods.

In-mold decoration can be done with either injection molding of thermoplastics or compression molding of thermosets. Thermosetting plastics are decorated with a two-stage process. For melamine products, for example, the mold is loaded with the molding powder in the usual manner and closed. It is opened after a partial cure, and the decorative "foil" or overlay is placed in position.

The mold is then closed again, and the curing cycle is completed. The overlay consists of a cellulose sheet having printed decoration and covered with a thin layer of partially cured clear melamine resin. During the molding cycle the overlay is fused to the product and becomes a part of the molding. The process is relatively inexpensive, especially when a multicolor decoration is required.

For in-mold decoration of thermoplastic products, single-stage process is used. The foil or overlay is thus placed in the mold cavity prior to the injection of the polymer. It is held in place in the mold by its inherent static charge. Shifting is prevented during molding inducing an additional charge by passing the wand of an electronic static charging unit over the foil after it is properly positioned.

The overlay, in all cases, is a printed or decorated film (0.003–0.005 in. thick) of the same polymer. Thus, polystyrene film is used for a polystyrene product, and polypropylene film for a polypropylene product. A similar procedure may also be used for decorating blow-molded products.

1.20.5 Embossing

Embossing is used for producing a tactile texture or pattern on plastics sheet or film. As the process involves the use of heat and pressure to texture a semifinished substrate, embossing is largely limited to thermoplastic materials. However, it can be adapted to thermoset composites, such as melamine-impregnated sheet stock.

Embossing is most commonly done with a two-roller system, in which one roller carries the embossing pattern and the other provides the essential pressure backup and feeding actions. Texture or pattern can be applied to the embossing or surface roller through a variety of processes, including conventional engraving, chemical engraving, etching, and laser cutting.

Embossing can also be performed without rollers, e.g., using textured aluminum foil in one-time use, or stainless steel plates with engraved textures can be used in the press cycle time and again, offering multiple impression economies.

Embossing is most frequently used as a method of decorating nonslip packaging materials, vinyl wall coverings, furniture laminates, building-panel laminates, textured foil for hot stamping, and other applications where the innate quality of three-dimensional printing is of value.

1.20.6 Electroplating

Electroplating is a chemical process for depositing heavy metals on plastics to achieve decorative effects and/or upgraded functionality. Since plastics are nonconductors of electricity, electroplating requires that the surface be properly conditioned and sensitized to receive metallic coatings. The principle of electroplating is to electrically conduct metal atoms such as copper, nickel and chrome off anodes placed within the plating baths through the plating solutions and onto the plastic production part. The target, i.e., the production part, acts as a cathode via connection to conductive plating racks, the part being attached to the plating rack with metal holding devices, spring-loaded contacts or prongs. The point of contact between the plating rack and the plastic part forms the continuity of the current flow from anode through the solution onto the plastic part.

The process of electroplating begins with the plastic part attached to the plating rack being subjected to preplate procedure, which is designed to create a surface on the plastic parts that will develop a bond between the plastic and the first nickel or copper deposit. These initial deposits are extremely thin, in the micron (10^{-6} mm) range. This first deposit is designed to increase conductivity uniformly over the plastic surface.

When preplating is completed (and the plastic articles have a conductive coating), it is possible to proceed to the electroplating operation, which is very similar to conventional electroplating on metal.

Electroplating of plastic products provides the high-quality appearance and wear resistance of metal combined with the light weight and corrosion resistance of plastics. Plating is done on many plastics, including phenolic, urea, ABS, acetal, and polycarbonate. Many automotive, appliance, and hardware uses of plated plastics include knobs, instrument cluster panels, bezel, speaker grilles, and nameplates. In marine searchlights zinc has been replaced by chrome-plated ABS plastics to gain lighter weight, greater corrosion resistance, and lower cost. An advantage of plastics plating is that, unlike metal die castings, which require buffing in most cases after plating, plastics do not ordinarily require this extra expensive operation. The use of plated plastics also affords the possibility of obtaining attractive texture contrasts.

1.20.7 Vacuum Metallizing

Vacuum metallizing is a process whereby a bright thin film of metal is deposited on the surface of a molded product or film under high vacuum. The metal may be gold, solver, or most generally, aluminum. The process produces a somewhat delicate surface compared to electroplating. The metallizing process can be used on virtually all properly (surface) prepared thermoplastic and thermosetting materials.

Small clips of the metal to be deposited are attached to a filament. When the filament is heated electrically, the clips melt and, through capillary action, coat the filament. An increased supply of electrical energy then causes vaporization of this metal coating, and plating of the plastic product takes place.

To minimize surface defects and enhance the adhesion of the metal coating, manufacturers initially give the plastics parts a lacquer base coat and dry in an oven. The lacquered parts are secured to a rack fitted with filaments, to which are fastened clips of metal to be vaporized. The vaporization and deposition are accomplished at high vacuum (about 1/2 micron). The axles supporting the part holding the fixtures are moved so as to rotate the parts during the plating cycle to promote uniform deposition. The thickness of the coating produced is about 5×10^{-6} in. (127 nm).

After the deposition is completed, the parts are removed and dipped or sprayed with a top-coat lacquer to protect the metal from abrasion. Color tones, such as gold, copper, and brass may be added to this coating if desired.

Vacuum metallizing of polymer films, such as cellulose acetate, butyrate, and Mylar, is performed in essentially the same way. Film rolls are unreeled and rewound during the deposition process to metallize the desired surface. A protective abrasion-resistant coating is then applied to the metallized surface in an automatic coating machine.

Vacuum metallizing is a versatile process used in a great variety of applications. Examples range from highly decorative cosmetic closures to automotive grilles and instrument clusters. Vacuum metallized plastic parts can replace metal parts with large saving in manufacturing costs and weight. The process can also serve functional needs, such as lamp reflectors or diffusion grids for overhead fluorescent lighting. Vacuum metallizing on interior surfaces of computer or communication equipment provides a degree of radio frequency interference shielding.

References

1. *Modern Plastics Encyclopedia*, Vols. 42–45, McGraw-Hill, New York, 1965–1967.
2. Bikales, N. M. ed. 1971. *Molding of Plastics*, p. 737, Interscience, New York.
3. Butler, J. A. 1964. *Compression Transfer Molding*, Plastics Institute Monograph, Iliffe, London.
4. Vaill, E. W. 1962. *Modern Plastics*, 40, 1A, Encycl. Issue, 767.
5. Maiocco, A. L. 1964. Transfer molding, past, present, and future, *SPE Tech. papers*, 10, XIV-4.
6. Frados, J. ed. 1976. *Plastics Engineeting Handbook, 4th Ed.*, Van Nostrand Reinhold, New York.
7. Rubin, I. I. 1973. *Injection Molding of Plastics*, Wiley, New York.
8. Brown, J. 1979. *Injection Molding of Plastic Components*, McGraw-Hill, New York.
9. Dyn, J. B. 1979. *Injection Molds and Molding*, Van Nostrand Reinhold, New York.
10. Lukov, L. J. 1963. Injection molding of thermosets, *SPE. J.*, 13, 10, 1057.
11. Morita, Y. 1966. Screw injection molding of thermosets, *SPE. Tech. Pap.*, 12, XIV-5.
12. O'Brien, J. C. 1976. Business injection molding thermosets, *Plast. Engl.*, 32, 2, 23.
13. Schenkel, G. 1966. *Plastics Extrusion Technology and Theory*, Elsevier, New York.
14. Bikales, N. M. 1971. *Extrusion and Other Plastics Operations*, Interscience, New York.
15. Fischer, E. G. 1976. *Extrusion of Plastics*, Newness-Butterworth, London.
16. Van Ness, R. T., De Hoff, G. R., and Bonner, R. M. 1968. *Mod. Plastics*, 45, 14A, Encycl. Issue, 672.
17. Fisher, E. G. 1971. *Blow Molding of Plastics*, Iliffe, London.
18. Elden, R. A. and Swan, A. D. 1971. *Calendering of Plastics*, Plastics Institute Monograph, Iliffe, London.
19. Mark, H. F., Atlas, S. M., and Cernia, E. eds. 1967. *Man-Made Fibers: Science and Technology*, Vol. 3, p. 91. Interscience, New York.
20. Moncrieff, R. W. 1963. *Man-Made Fibers*, Wiley, New York.
21. Riley, J. L. 1956. In *Polymer Processes*, C.E. Schildknecht, ed., p. 91. Interscience, New York, Chap. XVIII.
22. Carraher, C. E. 2002. *Polym. News*, 27, 3, 91.
23. Butzko, R. L. 1958. *Plastics Sheet Forming*, Van Nostrand Reinhold, New York.
24. Sarvetnick, H. A. ed. 1972. *Plastisols and Organosols*, p. 58, Van Nostrand Reinhold, New York.
25. Lubin, G. ed. 1982. *Handbook of Composites*, p. 58, Van Nostrand Reinhold, New York.
26. Lee, W. J., Seferis, J. C., and Bonner, D. C. 1986. Prepreg processing science, *SAMPE Q.*, 17, 2, 58.
27. Tsunoda, Y. 1966. U.S. Pat. 3,286,969.
28. Otani, S. 1965. *Carbon*, 3, 213.
29. Daumit, G. P. 1987. Latest in carbon fibers for advanced composites, *Performance Plastics' 87 First International Ryder Conference on Special Performance Plastics and Markets*, Feb., pp. 11–13. Atlanta, Georgia.
30. Becker, W. E. ed. 1979. *Reaction Injection Molding*, p. 368, Van Nostrand Reinhold, New York.
31. Kresta, J. E. ed. 1985. *Reaction Injection Molding*, ACS Symp. Ser., Vol. 270, p. 368, American Chemical Society, Washington, DC.
32. *PELASPAN Exapandable Polystyrene*, Form 171–414, Dow Chemical Co., 1966.
33. Klempner, D. and Frisch, K. C. eds. 1991. *Handbook of Polymeric Foams and Foam Technology*, p. 368, Hanser, Munich.
34. Zizlsperger, J., Statny, F., Beck, G., and Tatzel, H. 1970. U.S. Pat. 3504 068, (to BASF).

35. Phillips, L. N. and Parker, D. B. V. 1964. *Polyurethanes: Chemistry, Technology, and Properties*, Iliffe, London.
36. *One-Step Urethane Foams*, Bull. F40487, Union Carbide Corp., 1959.
37. Harris, T. G. 1981. U.S. Pat. 4281069 1981 10728 (to Armstrong World Industries, U.S.A.).
38. Chandra, P. and Kumar, A. 1979. Ind. Pat. 147107 19791117 (to Bakelite Hylam Ltd., India).
39. Lasman, H. R. 1967. *Mod. Plastics*, 45, 1A, Encycl. Issue, 368.
40. Gluck, D. G., Hagan, J. R., and Hipchen, D. E. 1980. In *Advances in Urethane Science and Technology*, Vol. 7, K.C. Frisch and D. Klempner, eds., p. 137, Technomic Publ. Co., Lancaster, Pennsylvania.
41. Schidrowitz, P. and Dawson, T. R. eds. 1952. *History of Rubber Industry*, p. 137, IRI, London.
42. Hoffman, W. 1967. *Vulcanization and Vulcanizing Agents*, Maclaren, London.
43. Nourry, A. ed. 1962. *Reclaimed rubber, Its developments, applications and Future*, p. 137 Maclaren, London.
44. Ghosh, P. 1990. *Polymer Science and Technology of Plastics and Rubbers*, Tata McGraw-Hill, New Delhi.
45. Kojima, M., Tosaka, M., Ikeda, Y., and Kohjiya, S. 2005. *J. Appl. Polym. Sci.*, 95, 137.
46. Parker, D. H. 1965. *Principles of Surface Coating Technology*, Wiley, New York.
47. O'Rinda Trauernicht, J. 1970. Bonding and joining plastics, *Plastics Technology, Reinhold Publishing, New York.*
48. Harris, R. M. ed. 1999. *Coloring Technology for Plastics*, Chemtec Publishing, Toronto.
49. *Petrothene: A Processing Guide*, 3rd Ed., 1965. U.S. Industrial Chemicals Co., New York.
50. Morgan, B. T., Peters, D. L., and Wilson, N. R. 1967. *Mod. Plastics*, 45, 1A, Encycl. Issue, 797.

2

Recycling of Polymers

2.1 Introduction

It is certainly true that plastics left lying around after use do not disappear from view and such post-consumer waste as foam cups, detergent bottles, and discarded film is a visual annoyance. This is because plastics are not naturally biodegradable. However, to consider this a detriment is a questionable argument. Rather, it may well be considered an advantage. This is borne out by the fact that recycling of plastics materials is now an important field in the plastics industry, not just an activity born under environmental pressure.

Although the plastics industry practiced recycling for many years, attention was mainly focused on the recycling of industrial scraps and homogeneous post-consumer plastics, which are easy to collect and reprocess. However, more recently the plastics industry accepted the challenge of recycling of heterogeneous plastics waste based on new technologies of separation and reprocessing. Scientific research, scarcely visible only a few years ago, is now a very active, fast-growing discipline, contributing to the development of newer processes.

According to the type of product obtained from the recycling process and the percentage of the economic value recovered, the following broad classification of recycling technologies can be made (1) primary recycling, the reprocessing of plastics waste into the same or similar types of product from which it has been generated; (2) secondary recycling, the processing of plastics wastes into plastics products with less demanding properties; (3) tertiary recycling, recovery of chemicals from waste plastics; and (4) quaternary recycling, recovery of energy from waste plastics.

The processes mainly used to these ends are: direct reuse after separation and/or modification, chemical treatment or pyrolysis for recovery of monomers and/or other products, and burning or incineration.

Primary recycling is used when the plastic waste is uniform and uncontaminated and can be processed as such. Only thermoplastic waste can be directly reprocessed; it can be used alone or, more often, added to virgin resin at various ratios. The main problems encountered in primary recycling are degradation of the material resulting in a loss of properties as appearance, mechanical strength, chemical resistance, and processability. Contamination of plastic scrap and handling of low-bulk density scrap such as film or foam are additional problems in primary recycling. Primary recycling is widely performed by plastics processors; it is often considered an avoidance of waste rather than recycling.

For post-consumer, mixed plastic wastes (MPW), which are unsuitable for direct use, the industry resorts to secondary recycling methods. There are various technical approaches to secondary recycling of MPW. These include reprocessing based on melt homogenization using specialized equipment; use of ground plastics waste as filler; and separation into single homogeneous fractions for further processing, such as partial substitution of virgin resins and blending with other thermoplastics using suitable compatibilizers.

In tertiary or chemical recycling of plastic wastes, polymers are chemically unzipped or thermally cracked in order to recover monomers or petrochemicals indistinguishable from virgin materials. Thermal cracking procedures offer viable alternatives by utilizing commingled plastics without decontamination. In quarternary recycling, energy content of plastics waste is recovered. In most cases, plastics are burned, mixed with other waste. Incineration of plastics alone creates a number of problems and requires the use of specially designed incinerators.

2.2 Outline of Recycling Methods

Post-consumer plastic wastes can be divided into two different groups depending on their source: (1) mixed plastics from the household waste and (2) plastics from the industrial sectors. The first category involves the medium-/short-life articles that are used in food, pharmaceutical, and detergent packaging, shopping, and others. The majority of these articles are composed of thin protective films: a variety of bottles for soft drinks, food, and cosmetics, sheeting for blisters, strapping and thermoformed trays.

There are basically five different polymers that contribute to the total amount of domestic plastic waste, namely, PE, PP, PS, PVC, and PET. The composition of this MPW can change depending on the regional habits and seasons of a year, and also on the mode of waste collection. A typical composition may be PE 39%, PVC 22%, PET 19%, PS 8%, and PP 12% (by wt).

The collection of plastics wastes always yields a polluted product, and this fact poses the need for the first operation of the recycling process, namely the cleaning of foreign bodies. The machinery required at this stage may be either manual or automatic type, the former being simpler from the standpoint of installation. The operations following the first step of clearing are determined by the type of recycling process to which the material is to be subjected. There are basically two main recycling processes: recycle of heterogeneous MPW and recycle of selected polymers separated from MPW.

A direct solution to disposal of domestic MPW can be the reuse of the heterogeneous mixture by processing through extrusion or injection molding technologies using traditional machineries. However, when MPW is processed, one of the main problems is to find the best compromise between homogenization and degradation. The optimal processing condition must ensure a good dispersion of the materials with high melting point (such as PET) in a continuous phase of molten polymers (such as PVC), avoiding gas bubbles, low-molecular-weight compounds, and cross-linked residues that are formed by thermal degradation. Some possible applications of such molded mixed plastics are injected tiles for paving, and extruded profiles for making structural articles such as benches, garden tables, bicycle racks, fences, and playing facilities. However, because of the incompatibility of the various components in mixed plastics, the mechanical properties of the molded or extruded products are rather poor.

The market of park benches, playgrounds, fences, and so on, cannot absorb, in the long run, the massive amounts of MPW that are produced every year. Hence the possible route to recycling of MPW to obtain secondary materials with acceptable mechanical properties could be to blend them with virgin polymers, or, at least, with recycled homopolymers. For example, experimental results [1] of processing and properties of blends of virgin LDPE and MPW have shown that all mechanical properties, with the exception of elongation at break, are very similar to those of the virgin material if the MPW content does not exceed 50%.

The possibility of using MPW as filler for both LDPE and HDPE has been considered [2] as such an approach, and may offer two important advantages: (1) improvement of the use of huge amounts of MPW that are generated by municipalities and industries; and (2) savings in nonrenewable raw materials and energy, both associated with the manufacturing of the virgin materials that can be replaced by plastics waste. Even if the percentage of plastics waste used as a filler cannot be higher, its common use may absorb sizable amounts of waste.

A widespread solution, in terms of application and market volume, can be the recycling of single materials or homogeneous fractions obtained from a differentiated collection system and/or a separation process of the mixture. Molded products from single or homogeneous fractions usually show a general performance far greater than that of products from mixed plastics. To obtain single or homogeneous fractions, it is useful to separate the mixed domestic plastics into four fractions, namely, polyolefins, PS, PVC, and PET.

An important preliminary to separation of mixed domestic plastics is the cleaning and selection operation. A simple method to perform this operation consists of a selection platform where a number of trained sorters separate the different types of plastics by visual assessment. Because manual selection is liable to human error, selection platforms may be equipped with detectors such as electronic devices to check the quality of the selected material.

The drawbacks of the manual platforms—which range from high labor cost to the complexity of labor management—may be avoided by resorting to automatic platforms. The machines required for such automation are manifold and the necessity to employ them is related to the quality of the collected material. Essential machines are rotary screen, light-parts separation equipment, heavy-parts separation equipment, and aluminum rejection equipment. All such machines are preliminary to the stage of separation into homogeneous plastic fraction.

Bottles constitute the largest high-volume component of post-consumer plastics and need special attention in reclaim operation. Since 1988, developments in bottle reclaim systems have made recycling post-consumer plastics more efficient and less costly. Municipalities, private organizations, universities, and entrepreneurs have worked closely to develop new collection, cleaning, and sorting technologies that are diverting larger portions of plastics from landfills to recycled resins and value-added end products.

To collect the high volume-to-weight ratio post-consumer plastics economically, truck-mounted compactors have been developed that seem to have the most promising future for mobile collection. They are self-contained and offer, on average, a reduction ratio of 10:1. Simple to operate, compactors accept all types of plastics, including film, and perform equally well with milk jugs and PET bottles as with mixed plastics.

Using compactors for on-board truck densification can thus be a cost-effective part of multimaterial collection programs. Another noteworthy development is that of flatteners and balers, which have also proven cost-effective under certain conditions. An integrated baler developed by Frontier Recycling Systems (USA) is fully automatic and capable of handling the plastic throughput of larger and costlier systems without the corresponding expenditures of space and labor. By producing smaller, high-density bales, it allows for lower transportation costs of recyclables shipped to market.

In keeping with the progress in densification options, efficient sortation systems have also been developed. The Poly-Sort integrated sorting line developed by Automation Industrial Control (AIC), Baltimore, is capable of sorting mixed stream of plastic bottles at a baseline rate of three bottles per second, or 700 kg/h. With expected advances in scanning and detection, the sorting rate of the system could double to 1400 kg/h.

Designed to sort compacted bottles, the Poly-Sort system employs conveyors for singulation, and two devices for color and chemical composition identification. A vibratory conveyor singulates bottles; a read conveyor transports bottles to an ultrasonic sensor that detects their position; a near-infrared system detects the resin type; a camera detects the color of the container; a computer integrates data and makes an identification; air jets divert bottles to the appropriate segregation conveyor or hopper.

The above type of separation is a macroseparation. It may be noted that the methods of separation into homogeneous fractions fall into three groups: molecular separation, microseparation, and macroseparation. Molecular separation is based on the dissolution of the various plastics in selective solvents, a method that is promising but still in the stage of study. Microseparation is a method by which a suspension medium is used to separate plastics with density higher or lower than the suspension medium. Macroseparation, which is the separation of plastics when waste materials are still in initial form, appears to be the most conveniently applicable system, considering the increasing possibilities of automation it offers. The key to this separation process is the development of an efficient detector system that can distinguish between type and quality of different plastics in waste materials.

Different types of detectors have been developed and many are under development. These are based on distinctive physicochemical properties of plastics and employ different techniques such as x-ray, near-infrared spectrophotometry, fluorescence, and optical measurement of transparency and color. Automatic systems consisting of a platform for selection according to plastics topology, a number of identification and detection steps, and adequate checks on the efficiency of separation following detection have been developed. The Poly-Sort system described above is one such example.

Recycle installations take up the separated plastic flakes for further processing. Various elements that normally compose the item to be recycled are caps made of PE, PE with PVC gaskets, aluminum, labels of tacky paper with different types of glue, and residues and dirt that have been added during the waste-collection phase. Various operations that are carried out in a specific sequence because of the problems posed by the type of material are: grinding to ensure homogeneity of the product, air flotation for separation of flakes with different specific weight and removal of parts of labels freed by grounding (such as separation of PVC labels from PET bottle flakes), and finally washing to remove residues. The washing system consisting of a range of equipment that includes centrifugal cleaners, washing tank, settling tank, and scraping machines is part of a know-how of various manufacturers.

The majority of municipal solid waste consists of plastics waste, which is often contaminated with significant amounts of paper. This is not only the case with plastics fraction of municipal solid waste (PFMW), but also with such industrial waste as used packaging materials, laminates, and trimmings. The reprocessing of plastics waste contaminated with more than 5% paper is difficult using conventional plastics processing machinery, and becomes almost impossible at paper levels exceeding 15%. The sorting operation at a municipal plant normally aims at removing the paper component from the light plastics fraction to a level well below 1%. However, this operation has not been quite successful because the material handling side has been difficult and the costs have far exceeded the price of virgin polyolefins.

A simpler solution to the problem of contamination may be to allow for a paper component in the plastics fraction and to use a processing method that can disintegrate the cellulose fibers into small fragments such that they act as particulate fillers in the plastics. Such a method has been developed at Chalmers University of Technology, Gothenburg, Sweden. Known as the CUT-method, the process makes it possible to reprocess both the PFMW and a number of different industrial plastic waste materials contaminated with paper [3,4]. The CUT-method, consisting of a prehydrolytic treatment of the paper component, is an industrially applicable method of reprocessing paper-contaminated plastics waste of various origins.

The main advantage of the CUT-method is that the plastics fraction and the paper component do not need to be separated and the hydrolysis does not degrade the plastics component but reduces the chain length of the cellulose component to a level at which the cellulose fiber becomes extremely brittle and the shear forces generated in normal plastics processing machinery (compounding extruders and molding machines) can easily disintegrate the paper parts into small fibrous fragments. It is the disintegration of the embrittled paper component into an almost pulverized substance that is the key to the success of the method, since this results in greatly enhanced melt flow properties, better homogeneity, and thus in improvement in the mechanical properties of the material [5].

The method of hydrolysis used in the CUT-method offers an efficient and economical way of processing plastic waste, both post-consumer municipal waste and industrial waste, contaminated with a cellulose component. The presence of cellulose gives a desired stiffness to the final product, as studies have shown [4,5]. Such plastics product can be used in several applications, such as artificial wood.

Plastics wastes from industrial sectors concern mostly the medium-/long-life articles, as plastics have played a fundamental role in the exceptional growth of production technology seen during recent years in these sectors, and in particular the automotive industry. Because of the advantage in design and functionality, plastics are now an indispensable part of any kind of car; the amount of polymers employed to build a car has risen to about 20% from a mere 5% in 1973, with a corresponding increase in the quantity of nonmetallic waste during scrapping. The main problem of plastic wastes from all industrial sectors, and in particular the car industry, is the large variety of materials employed to build a single component or system, for example, a dashboard. This takes place because of the sophisticated and complex mission that the system must perform. The large number of plastics used and the disproportionately high costs in the dismounting of the different plastic pieces of a car represent an intractable waste-recovery problem and thus have a negative impact on the recycling process. As a result of this, only the metallic fraction is recovered, while the plastic materials continue to be eliminated by deposition in refuse dumps.

An alternative approach to the recovery of automotive plastics is therefore to use them as large, easily removable components that offer potential for reclamation as well-characterized individual polymers. Some particularly complex components such as vehicle front- and rear-end systems, exhibit special suitability for manufacture in plastics instead of metals because of their ease of production and assembly. It is generally recognized that improvements in automotive scrapyard economics may be best achieved by the prior removal from vehicles of such large polymeric components and their recycling as well-characterized plastic fractions. For example, plastic fuel tanks of HDPE are now in common use and represent the most common recyclable plastic component. Trials with material recovered from used plastic fuel tanks have shown promising results for the manufacture of new tanks [6].

A concept that is being developed to solve the recycling problems of plastics from industrial sectors, and in particular the car industry, is based on the use of materials of the same family for all components of the plastic systems to be recycled at the life end. This allows an easy and direct recycling of the scraps and the recovery of the whole system. Greater recycling efficiency can be obtained when the following two basic requirements are satisfied: (1) materials compatibility through materials homogeneity, and (2) easier disassembly through planned design. This concept has been first applied to the automotive sector, where the environmental problems have become of primary importance; however, it could be also applied to other products, i.e., appliances and building materials. There are two tasks in developing this concept: to develop new advanced materials in individual categories of polymers and to promote new technologies.

Consider, for example, the automotive industry. Although many polymers are used in cars today, the industry tends to favor more and more polypropylene use due to a large range of properties available. New developments in polyolefin-based materials have thus created a family of polypropylene products with a wide range of physical properties, including the ability to be easily recycled. When utilized by automotive and product designers as a part of a design for disassembly strategy, these compatible materials will yield large subassemblies that can be reclaimed with a minimum of handling [7,8]. In each project, the design incorporates readily identifiable hard point connections between the polypropylene components and the metal automobile subframe. This allows personnel in recycling centers to remove these parts quickly and in large pieces that can be completely reground and recycled. This concept has been applied to car dashboards and interior vehicle components like floor covering, trim, and door panels, as well as bumpers.

Blends of EPDM rubbers with polypropylene in suitable ratios have been marketed as thermoplastic elastomers (TPE), also commercially known as thermoplastic polyolefin elastomers (TPO). These heterophasic polymers, characterized by thermoreversible interaction among the polymeric chains, belong to a broad family of olefinic alloys that can now be produced directly during the polymerization phase, unlike blended TPE and TPO, and various compositions (with various compounding additives) can be formulated which are primarily tailored to meet different requirements of most of car applications. The TPE-based synthetic leather and foam sheets are typical examples.

In order to obtain all-TPE recyclable applications, different assembly techniques have been specifically studied to obtain the basic composite structures [8,9]. The most interesting technique is one that allows simultaneously thermoforming, embossing, and coupling to be obtained in one stage of operation, yielding a foamed synthetic leather bilayer on a rigid support (all TPE based) without adhesives. With new designs for recycling, dashboard, floor covering, and other interior components such as door panels, pillar trim, and rear shelf have been made of the same chemical material (polypropylene) in different forms, thus providing an important aid to the recycling of plastic.

Lead-acid batteries from automotive applications normally have a shorter service life than the car itself. The logistics system for used car batteries is geared to lead recycling. However, the first battery reprocessing step yields not only lead but also polypropylene in a form of the casing fragments. Accordingly, the polymer is available without additional cost. As the casing makes up a substantial part (7%) of the total battery, the quantities of polypropylene obtained are sufficient to warrant the operation of a plastics recycling plant. For example, BSB Recycling GmbH in Braubach, Germany, a subsidiary of Metallgesellschaft AG, operates secondary lead smelter for lead recovery from postuse lead-acid batteries [10]. They process some 60,000 tons of batteries per annum, which accounts for half of the used battery volume to be disposed in the western states of Germany. BSB started to segregate the polypropylene from the battery casings and route it to a separate recycling process as far back as 1984. For the recycling process, a quality assurance system geared to the specific requirements of the applications has been developed and implemented.

Polyolefins and poly (ethylene terephthalate) (PET) are the most frequently recycled polymers obtained from both the domestic and industrial plastics wastes, and as such they have received most attention in the recycling research and technology. PET is one of the largest recycled polymers by volume [11], because it is suitable for practically all recycling methods [12]. Over 50% of the PET film produced in the world is used as a photographic film base. The manufacturers of these materials have long been interested in PET film recovery. An important motivation for this has been the fact that photographic films are usually coated with one or more layers containing some amount of rather expensive silver derivatives.

Silver recovery makes PET-base recovery more economical. In a typical way of operation, PET film recycling is thus coupled with the simultaneous recovery of silver, for example, by washing with NaOH and follow-up treatment. PET-recycling by direct reuse, if the washed PET-film scrap is clean enough to be recovered by direct reextrusion, is by far the most economical process. However, this process is most suited for the recovery of in-production wastes. For customer-recollected PET-film, which may have a higher degree of contamination, other technologies are to be applied.

There exists a hierarchy in PET-film and plastics recycling technologies depending, first of all, on the degree of purity of PET scrap to be handled, and secondly, the economics of the process. For the cleanest PET grade, the most economical process, i.e., direct reuse in extrusion, is self-explanatory. For less-clean PET waste, it is possible to reuse them after a modification step (partial degradation, e.g., by glycolysis) at a reasonably low price. More-contaminated PET waste must be degraded into the starting monomers, which can be separated and repolymerized afterwards, of course at a higher cost.

Polyethylene films from greenhouses, although highly degraded by UV radiation, are recycled by various means leading to manufacture of films and molded products with low mechanical properties. Problems in the recycling of greenhouse films arise form the presence of products of photooxidation, which significantly affect the properties of a recycled material. An interesting possibility of use of photooxidized PE in blends with nylon-6 to improve blend compatibility has been demonstrated [13,14]. These follow from the earlier efforts [15,16] to compatibilize blends of polyamides and polyolefins (which are potentially very interesting, but, because of the strong incompatibility of both polymers, yield products having poor properties) with the use of functionalized polyolefins that can react with the amino groups of polyamides, giving rise to copolymers and thus stabilizing the blend.

Such functionalization, in general a long and extensive step, is mostly performed by chemical modification of the polyolefin structure. However, studies have demonstrated [13] that photooxidized PE offer similar results. Thus the use of recycled (photooxidized) greenhouse PE in blends with nylon give rise to PE/nylon graft copolymers during processing, which improve the mechanical properties of the resultant material. The graft copolymers act as compatibilizing agents; the properties of nylon-rich blends (80 wt% nylon-6) thus are found to be very similar to those of blends compatibilized by PE, and which is initially functionalized by chemical means. Moreover, in coextrusion, a good adhesion between the two layers (nylon and recycled PE) of coextruded films helps to avoid a need for the addition of a third layer binding two incompatible phases.

Chemical means such as glycolysis, methanolysis, and hydrolysis are good at unzipping only the condensation polymers—such as polyester, nylon, and polyurethanes—to facilitate chemical recycling.

Addition polymers, such as vinyls, acrylics, fluoroplastics, and polyolefins, can hardly be reprocessed except that, if they are sorted, they may be converted into powder by grinding operation and mixed with respective virgin resins for remolding into finished goods or, in some cases, blended with other resins using suitable compatibilizers to make useful end-products of commercial value.

Tertiary recycling of addition polymers require pyrolysis, which is a more aggressive approach. For mixed or unsorted plastics in particular, it is a practicable way of recycling. Pyrolysis is the thermal degradation of macromolecules in the presence of air. The process simultaneously generates oils and gases that are suited for chemical utilization.

The advantage of pyrolysis over combustion (quaternary recycling) is a reduction in the volume of product gases by a factor of 5–20, which leads to considerable savings in the gas conditioning equipment. Furthermore, the pollutants are concentrated in a coke-like residue matrix. It is possible to obtain hydrocarbon compounds as gas or oil.

The pyrolysis is complicated by the fact that plastics show poor thermal conductivity, while the degradation of macromolecules requires considerable amounts of energy. The pyrolysis of mixed plastic wastes and used tires has been studied in melting vessels, blast furnaces, autoclaves, tube reactors, rotary kilns, cooking chambers, and fluidized bed reactors [17,18].

Rotary-kiln processes are particularly numerous. They require relatively long residence times (20 min or more) of the solid wastes in the reactor. Moreover, due to the large temperature gradient inside the rotary kiln, the product spectrum is very diverse. For this reason, the gases and oils generated by the pyrolysis are normally used for the direct generation of energy and the process may well be considered as a quaternary recycling process.

For chemical recycling of mixed plastics, the fluidized bed pyrolysis has turned out to be particularly advantageous. The fluidized bed is characterized by an excellent heat and mass transfer as well as constant temperature throughout the reactor. This results in small dwell times (a few seconds to a 1.5 min maximum) [18] and largely uniform product spectra. The fluidized bed is generated by a flow of air or an inert gas (nitrogen) from below through a layer of fine-grained material, e.g., sand or carbon black. The flow rate is sufficient to create turbulent motion of particles within the bed. Using a fluidized bed pyrolysis, 25–45% of product gas with a high heating value and 30–50% of an oil rich in aromatics could be recovered [18]. The oil is comparable to that of a mixture of light benzene and bituminous coal tar. Up to 60% ethylene and propylene are produced by using mixed polyolefins as feedstock. Moreover, depending on the temperature and the kind of fluidizing gas (nitrogen, pyrolysis gas, and steam) different variants of the fluidized bed pyrolysis process can be carried out, yielding only monomers, BTX-aromatics, high boiling oil, or gas.

A promising concept that is receiving increasing attention is recycling plastics to refinery cokes, where pyrolysis units and a well-developed infrastructure are already in place. The main hindrance to the execution of this concept is the presence of contaminants (including chlorine and nitrogen) in the plastics stream, as well as the need to turn plastics into a liquid form that the refinery can handle. Projects are in place to address these issues. Initial small-scale pyrolysis, dissolving plastics into other refinery feedstocks, or turning solid wastes into a slurry, are some of the options that have received attention. Efforts have also been made in some refineries to convert mixed plastics into a petrochemical feedstock by catalytic hydrogenation. In the refinery, the aim of tertiary recycling is not to displace regular refinery capacity, but to use plastic waste as a very minor stream. However, even if all refineries with cokers took only 2% of their capacity as plastic waste, it would be extremely significant.

Mention should be made of a plastics liquefaction process that has been developed jointly by the Japanese Government Industrial Laboratory, Hokkaido, Mobil Oil Corporation, and Fuji Recycle Industry. The process can treat polyolefinic plastics (polyethylene, polypropylene, and polystyrene) or their mixtures by a combination of thermal and catalytic cracking to produce gasoline, kerosene, and gas oil fractions of about 85%. Recovered liquid and gas are separated by cooling and the gas is used as in-house fuel. The technology is unique in using proprietary Mobil ZSM-5 catalyst and has been described as an ultimate recycling technology [19].

A brief overview of several important aspects of plastics recycling and development in the field has been given above. Some of the topics that have been highlighted in this review will now be elaborated further in the following sections. In addition, waste recycling problems and possibilities relating to a number of common plastics will be discussed.

2.3 Recycling of Poly(Ethylene Terephthalate)

The largest use of poly (ethylene terephthalate) (PET) is in the fiber sector, with PET film and PET bottles representing only about 10% each of the total PET volume produced annually. A large percentage of the total PET output comprising films, plastics, and fibers is recycled by various methods and for several applications, which makes PET one of the largest in volume of recycled polymers in the world. Contributing to this is the suitability of PET for practically all recycling methods, which include direct reuse, reuse after modification, recovery of monomers and other low-molecular-weight intermediates, and incineration. Any particular method is selected on the basis of the quality of waste and scrap, the economy of the process, and the convenience of the operation.

Contamination of post-consumer PET (POSTC-PET) is the major cause of deterioration of its physical and chemical properties during reprocessing. POSTC-PET is contaminated with many substances: (1) acid producing contaminants, such as poly(vinyl acetate) and PVC; (2) water; (3) coloring contaminants; (4) acetaldehyde; (5) other contaminants such as detergents, fuel, pesticides, etc., stored in PET bottles.

The most harmful acid to the POSTC-PET recycling process are acetic acid, which is produced by poly (vinyl acetate) closures degradation, and hydrochloric acid produced by the degradation of PVC. The acids act as catalysts for the chain scission reactions during POSTC-PET melt processing. Thus, the presence of PVC, as little as 100 ppm, would increase POSTC-PET chain scission [20]. Water reduces molecular weight (MW) during POSTC-PET recycling through hydrolysis reactions at the processing temperature (280°C). Moisture contaminants should be below 0.02% to avoid such MW reduction [21].

Acetaldehyde is present in PET and POSTC-PET, as it is a by-product of PET degradation reactions. The migration of acetaldehyde into food products from PET containers was a major concern in the early stages of developing the recycling process. Acetaldehyde being highly volatile, it can be minimized by processing under vacuum or by drying. Stabilizers such as 4-aminobenzoic acid, diphenylamine, and 4,5-dihydroxybenzoic acid are added to PET in order to minimize the generation of acetaldehyde [22].

2.3.1 Direct Reuse

This method, also called recycling by re-extrusion or melt recovery, is used for relatively pure PET waste such as cleaned consumer bottles or in-house waste. The method is based on the same principles as the original equilibrium polycondensation reaction:

$$
\begin{aligned}
&H{-}\!\left[ET\right]_m{-}OCH_2CH_2OH + H{-}\!\left[ET\right]_n{-}OCH_2CH_2OH \rightleftharpoons \\
&\qquad H{-}\!\left[ET\right]_{m+n}{-}OCH_2CH_2OH + EG \\
&[ET] \equiv \left[{-}OCH_2CH_2OOC{-}C_6H_4{-}CO{-}\right] \quad \text{(ethylene terephthalate)} \\
&EG \equiv HOCH_2CH_2OH \quad \text{(ethylene glycol)}
\end{aligned}
\tag{2.1}
$$

As polymer buildup and polymer degradation are taking place in the melt simultaneously, the reaction conditions have to be controlled very carefully in order to obtain the desired molecular weight and molecular weight distribution for the end use. In theory, this seems rather simple; in practice, however, a large amount of determining parameters (temperature, environmental atmosphere, holding time in a melt state, amount of impurities, type of used catalysts, stabilizers, etc.) have to be kept under control.

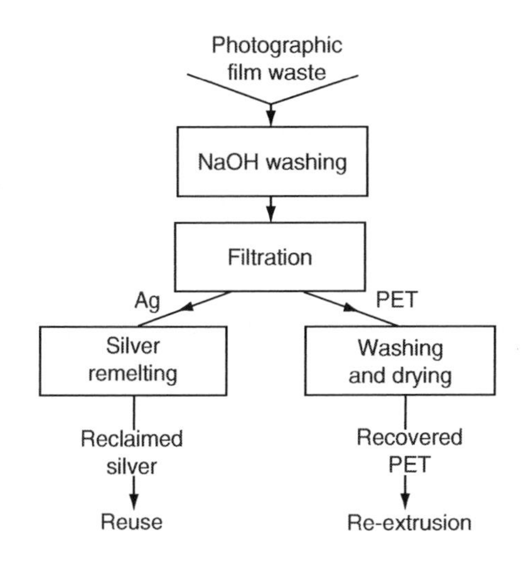

FIGURE 2.1 Combined recovery of silver and PET. (After De Winter, W., *Die Makromol. Chem., Macromol. Symp.*, 57, 253, 1992. With permission.)

A practicable reextrusion process was worked out and described by Syntex Chemie nearly forty 30 years ago [23]. This method—with some modifications—is still being used. The greatest recycler of fiber waste in the U.S. is Wellman; they recover PET fiber and bottle waste for home furnishing and nonwoven materials by a similar method.

Customer-recollected waste from fiber and textiles consists mainly of continuous filaments or staple fibers, which may be contaminated with dyestuffs, finishes and knitting oils, and other fibers such as cotton, wool, rayon, nylons, and acrylics; they are the most difficult-to-recover products.

A different picture can be presented for the PET bottles. In the environmentally active states in the U.S., 80–95% of the PET bottles sold are recollected and recycled. In Europe and in Japan where recycling has started earlier than in the U.S., various reclamation and reprocessing methods have been worked out and applied in practice. Because these processes are usually proprietary, the details of their operation are not known.

The larger use of PET film is as a photographic film base, which accounts for over 50% of the PET film produced in the world. The manufacturers of these materials, mainly Agfa-Gaevert, Eastman Kodak, du Pont de Nemours, Fuji, 3M, and Konishiroku, have long been interested in the recovery of PET film because of its content of rather expensive silver derivatives. Recycling of PET-film waste in production, which may amount to 25–30% of the total output, is almost complete by these manufacturers.

In a typical way of operation, PET film recycling is coupled with the simultaneous recovery of silver, as represented schematically in Figure 2.1. In the first step of the process, photographic emulsion layers containing silver are washed with, for example, NaOH, and after separation, silver is recovered on one side and cleaned PET waste on the other side [24]. Careful analysis is necessary to ensure that the washed PET-film scrap is clean enough to be recovered by direct extrusion.

The most obvious way of adding the recycled PET flakes is after the usually continuous polymerization and before the PET melt enters the extruder screw [25]. Such a procedure, however, has two main drawbacks: first, the highly viscous melt is difficult to filter (to eliminate possible gels or microgels); and second, other impurities (e.g., volatiles, oligomers, and colored parts) cannot be eliminated any more. In order to remove these disadvantages, several alternative modes have been worked out. A method to add recycled PET to the polymerization batch reactor during the esterification step was described by du Pont as early as 1960 [26]. Such a method shows the following advantages over the method described above: filtration can take place in the low-viscosity phase, and volatiles can be eliminated during the prepolymerization phase.

PET recycling by direct reuse, as described above, is by far the most economical process. However, it is useful in practice only for well-characterized PET wastes that have exactly known chemical composition (catalysts, stabilizers, and impurities). Therefore, the method is ideally suited for the recovery of in-production wastes, but it may not be suitable for post-consumer PET film.

2.3.2 Reuse after Modification

For post-consumer PET waste having a higher degree of contamination, technological processes based on degradation by either glycolysis, methanolysis, or hydrolysis can be used. These yield products that can be

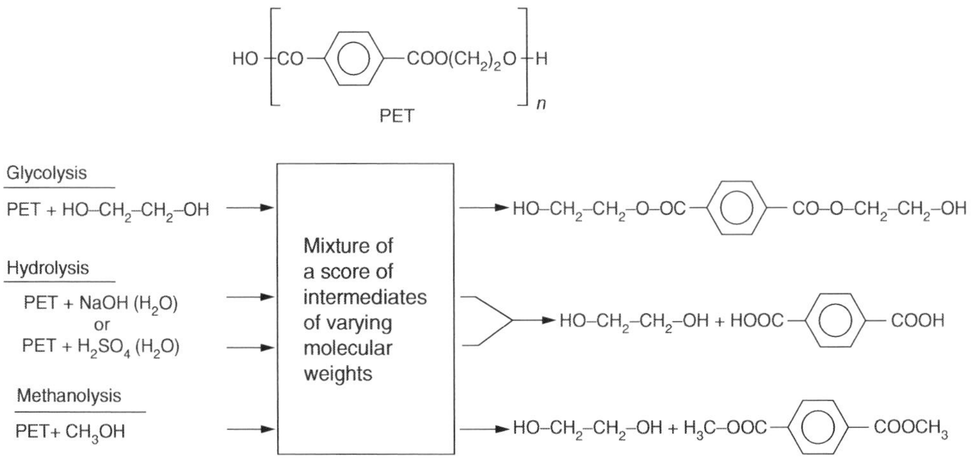

FIGURE 2.2 PET degradation by glycolysis, hydrolysis, and methanolysis. (After De Winter, W. 1992. *Die Makromol. Chem., Macromol. Symp.*, 57, 253.)

isolated. The principles of chemical processes involved in these methods are schematically represented in Figure 2.2.

Hydrolysis and methanolysis of PET regenerates the starting monomers. Thus, terephthalic acid (TPA) along with ethylene glycol (EG) are obtained by hydrolysis, while methanolysis yields EG and dimethyl terephthalate (DMT) among other products. Stopping short of complete depolymerization, glycolysis degrades long polymer chains (with typical repeat sequences of 150 units) into short-chain oligomers (repeat sequences of 2–10 units) having hydroethyl end groups.

2.3.2.1 Glycolysis

The addition of EG–PET reverses the polymerization reaction. This can be stoichiometrically represented by

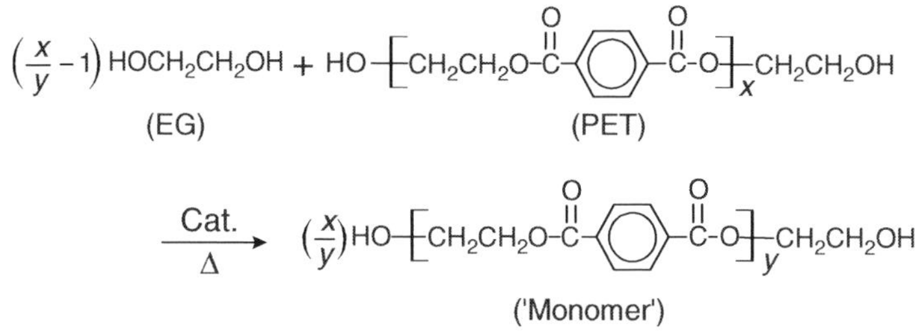

$$\left(\frac{x}{y}-1\right)\underset{\text{(EG)}}{HOCH_2CH_2OH} + \underset{\text{(PET)}}{HO\!\left[CH_2CH_2O\text{-}\overset{O}{\overset{\|}{C}}\text{-}C_6H_4\text{-}\overset{O}{\overset{\|}{C}}\text{-}O\right]_x\!CH_2CH_2OH} \xrightarrow[\Delta]{\text{Cat.}} \left(\frac{x}{y}\right)\underset{\text{('Monomer')}}{HO\!\left[CH_2CH_2O\text{-}\overset{O}{\overset{\|}{C}}\text{-}C_6H_4\text{-}\overset{O}{\overset{\|}{C}}\text{-}O\right]_y\!CH_2CH_2OH} \tag{2.2}$$

where x = average number of repeat units in polymer and y = average number of repeat units in 'monomer.' When $y = 1$, monomer = dihydroxyethyl terephthalate (DHET).

Glycolysis thus represents a compromise between regeneration of starting ingredients by methanolysis or hydrolysis and direct melt recovery. It is less costly than the former and more versatile than the latter. The resultant, easily filtered, low viscosity 'monomer' can be repolymerized to a useful higher molecular weight product. A typical flow sheet of the process is shown in Figure 2.3.

PET scrap suitable for glycolytic recycle includes production waste, fibers, film, flake, and bottles. In a practical system, major contaminants are separated from feedstocks, e.g., bottle waste is cleaned and separated from a polyethylene base, paper labels, metallic caps, and liners. For many end uses, colored PET must also be segregated. (Highly modified copolymers, glass-reinforced resin, fiber, or fabric blends are not suitable for glycolysis. These can only be recovered by methanolysis/hydrolysis.) Since reaction time depends on surface area, PET feedstocks must be reduced to relatively small particles by grinding, cutting, etc.

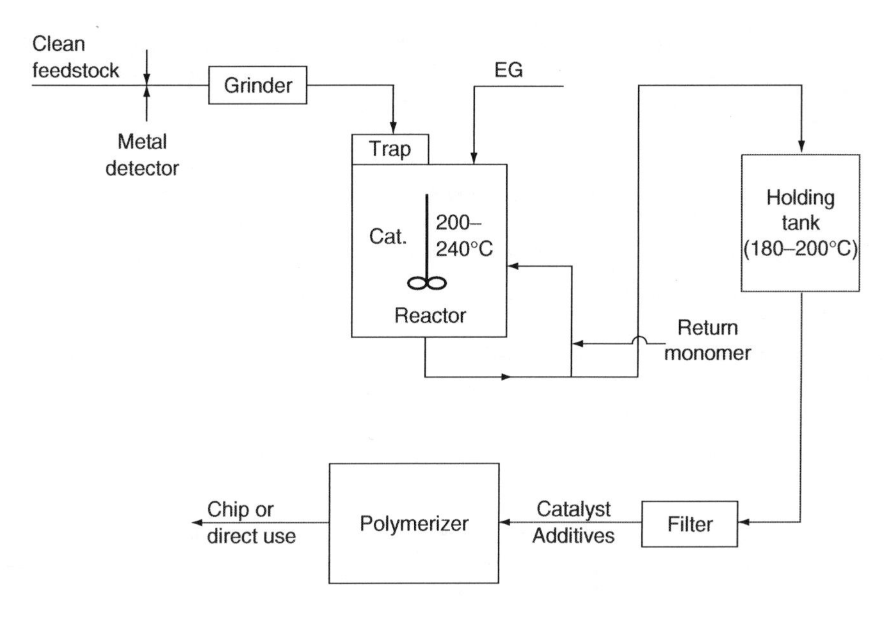

FIGURE 2.3 Flow diagram of a typical system for glycolytic recycling of PET waste. (After Richard, R., *ACS Polym. Prepr.*, 32(2), 144, 1991. With permission.)

The du Pont company [27] published many details covering the glycolytic recycling of PET. Goodyear has also developed a PET recycling process based on glycolysis that is called REPETE [28]. In a batch process, a molten 'monomer' heel is left in the reactor to allow the feedstock/glycol mixture to reach optimum reaction temperatures. In a continuous process (Figure 2.3) some of the molten 'monomer' is recycled to a stirred reactor to accomplish the same function. High glycol/terephthalate (G/T) ratios lead to more complete glycolysis but lower the maximum temperature, increasing the reaction time. A ratio of 1.7–2.0 G/T is a practical compromise [29]. An ester exchange catalyst as zinc or lithium acetate is usually added to increase the rate of glycolysis. Reaction temperatures of 220–240°C and times of 60–90 min are typical. The reactor is operated under a positive pressure to prevent forming an explosive mixture of air and glycol vapors.

The major side reaction is the production of ethers:

$$
\begin{aligned}
&2HOCH_2CH_2OH \xrightarrow{\text{Acid}} HOCH_2CH_2OCH_2CH_2OH \longrightarrow \\
&\qquad HO\!-\!\!\left[ET\right]_n\!-CH_2CH_2O\!-\!\!\left[ET\right]_m\!-OCH_2CH_2OH \\
&ET \equiv \left[OCH_2CH_2OOC-C_6H_4-CO \right]
\end{aligned}
\tag{2.3}
$$

Since this reaction is acid catalyzed, it can be minimized by adding a buffer such as sodium acetate or by adding water [30]. Lithium acetate catalyst also produces less ethers than since acetate. Some other side reactions are the formation of aldehyde, cyclic trimer of ET, and dioxane. Oxidation of glycol ends produce aldehydes that lead to colored compounds. Traces of dioxane can form from the cyclization of glycol.

If other glycols, such as diethylene glycol, are substituted for ethylene glycol, the corresponding oligomers are formed. These can subsequently be polymerized with aliphatic diacids as adipic or 4,4′-diphenylmethane diisocyanate to give rigid elastomers [31]. Additive to control luster, color, and so on can be added in the usual manner before and after polymerization.

Primary uses for PET from glycolytic recycle are geotextiles, fibers for filling products, nonwovens, and molding resins where color, strength, and control of dyeability is not important. Recovered polymer can be added to virgin polymer for films, fibers, and molding resins.

2.3.2.2 Methanolysis

PET waste obtained in the form of film, bottles, and fibers can be very conveniently converted into its raw materials dimethyl terephthalate (DMT) and ethylene glycol (EG) by methanolysis. The process involves heating the PET waste with methanol at 240–250°C and 20–25 kg/cm^2 pressure in the presence of catalysts such as metal oxalates and tartrates. Once the reaction is completed, DMT is recrystallized from the EG-methanol molten liquor, and distilled to obtain polymerization-grade DMT. Also EG and methanol are purified by distillation. Eastman Kodak has been using such a process for recycling of x-ray films for nearly 40 years and it is still improving the process [32], e.g., by using superheated methanol vapor to allow the use of ever more impure PET waste. Important factors that have to be dealt with in this process are avoiding coloration due to aldehyde formation and minimizing the formation of either glycols.

2.3.2.3 Ammonolysis

PET wastes can be converted via ammonolysis to paraphenylenediamine, which is a basic raw material for the high-modulus-fiber Kevlar or for high-value hair dyes. The chemical basis for this process is a modified Hoffman rearrangement. The synthesis may be done via the following three stages [33]:

Step 1

$$\left[-OCH_2CH_2OC(=O)-C_6H_4-C(=O)- \right]_n \xrightarrow{NH_3} H_2NC(=O)-C_6H_4-C(=O)NH_2 + HOCH_2CH_2OH$$

(I)

Step 2

$$H_2NC(=O)-C_6H_4-C(=O)NH_2 \xrightarrow{Cl_2} ClHNC(=O)-C_6H_4-C(=O)NHCl + HCl$$

(II)

Step 3

$$ClHNC(=O)-C_6H_4-C(=O)NHCl \xrightarrow{NaOH} H_2N-C_6H_4-NH_2 + NaCl + Na_2CO_3$$

(III)

In the first step, granulated PET is suspended in ethylene glycol and treated with gaseous ammonia at 100–140°C. In this reaction, the ethylene glycol also acts as a catalyst. The product terephthalimide (I) is insoluble in the medium and thus may be isolated. In the second step, terephthalimide (I) is suspended in water and chlorinated vigorously with chlorine gas. The resulting terephthalic bis-chloramide (II) is treated with NaOH solution to obtain paraphenylene diamine (III). An important aspect of this process is that paraphenylenediamine so obtained is completely free from its ortho and meta isomers and its production cost is much less than the market price. ICI has reported an alternative single-step process for conversion of PET to paraphenylenediamine by ammonolysis in the presence of hydrogen gas.

2.3.2.4 Hydrolysis

PET can be completely hydrolyzed by water at higher temperatures and pressure in the presence of catalysts (acidic as well as alkaline) to regenerate the monomers, terephthalic acid, and ethylene glycol.

While both acid- and base-catalyzed systems are completely realistic, their usefulness under practical production conditions remain controversial. As far as acid hydrolysis is concerned, the large acid consumption and the rigorous requirements of corrosion resistance of the equipment make profitability questionable. Moreover, the simultaneous recovery of TPA and EG, requiring the use of ecologically undesirable halogenated solvents, is difficult and not economical. For the alkaline hydrolysis process, also, the profitability is strongly determined by the necessity of expensive filtration and precipitation steps. In spite of the fact that the majority of newer industrial PET-synthesis plants are based on the TPA process rather than on the DMT process [34], the hydrolytic method of PET recycling has not gained favor.

2.3.2.5 Depolymerization in Supercritical Fluids

The supercritical fluid over its critical point has high density, such as in liquid state, and high kinetic energy as in a gas molecule. Therefore the reaction rate is expected to be higher than the reaction under liquid state conditions. PET is depolymerized quickly by solvolysis in supercritical water [35] or supercritical methanol [36]. The main products of PET depolymerization in supercritical methanol are dimethyl terephthalate (DMT) and ethylene glycol (EG), as shown in Figure 2.4. The depolymerization is carried out typically at temperatures between 543 and 603 K under pressures of 0.1–15 MPa for a reaction time of 3–60 min. For example, at 573 K, sample/methanol ratio 1/5 (by wt) and reaction pressure 14.7 MPa, DMT yield is reported [37] to be 98 per cent in 30 min.

It has been suggested that random scission of polymer chain takes place predominantly in the heterogeneous phase during the initial stage of PET depolymerization in supercritical methanol producing oligomers, whereas specific (chain end) scission to monomers proceeds predominantly in the homogeneous phase during the final stage.

2.3.2.6 Enzymatic Depolymerization

In 1977, Tokiwa and Suzuki reported that some lipases, which are extracellular enzymes that usually cleave esters in oils and fats, are also able to attack ester bonds in some aliphatic polyesters and can depolymerize such materials [38]. Aliphatic polyesters, however, exhibit only limited useful properties for many applications. Aromatic polyesters, such as PET and PBT, which are widely applied because of their excellent properties, are not attacked by hydrolytic enzymes. This led to the development of aliphatic-aromatic polyesters as biodegradable plastics that present a compromise between biodegradability and material properties [39]. Recently, however, Müller et al. [40] have isolated a hydrolase (TfH) from *Thermofibida fusca* which is able to depolymerize the aromatic polyester PET at a high rate in contrast to other hydrolases such as lipases. They have demonstrated for the first time that commercial PET can be effectively hydrolyzed by an enzyme at a rate that does not exclude a biological recycling of PET. The effective depolymerization of PET with the enzyme TfH will result in water

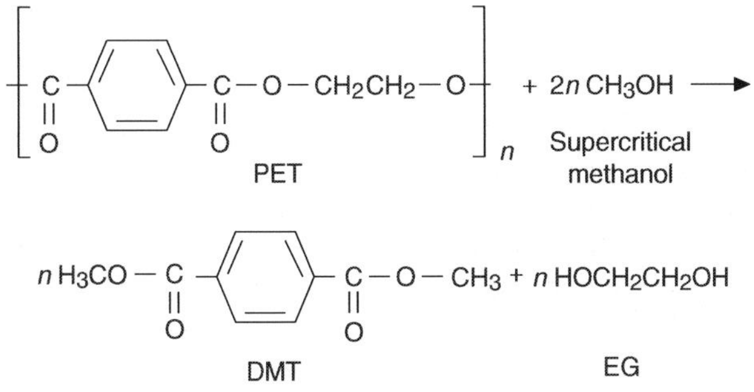

FIGURE 2.4 Main reaction of PET depolymerization in supercritical methanol.

soluble oligomers and/or monomers that can be reused for synthesis. In contrast, a microbial treatment of PET may not be appropriate for recycling purposes, since monomeric and oligomeric depolymerization products would be consumed by the microorganisms involved or inhibit their action and growth [40].

It is likely that the degradability of PET with hydrolases such as TfH strongly depends on the polymer crystallinity and the temperature at which the enzymatic degradation takes place [40]. The effective enzymatic PET hydrolysis will thus be expected to occur only below a certain critical degree of crystallinity. However, for bottle manufacture polyesters with low crystallinity are preferred for high transparency, thus increasing the susceptibility of PET to enzymatic attack.

One reason for the high activity of TfH hydrolase towards PET may be the high temperature (55°C) optimum, which is a result of its origin from a thermophilic microorganism. However, differences in the degradation behavior between TfH and the other lipases may also be due to differences in the structure of the enzymes, possibly enabling TfH to attack less mobile polyester segments and degrade PET at a surprisingly high rate.

2.3.3 Incineration

For PET wastes containing a large amount of impurities and other combustible solids it is more profitable to resort to quaternary recycling, that is, energy recovery by burning. Research along this line has been performed, particularly in Europe and Japan, since the early 1960s. Strong emphasis has been laid on the optimization of incinerators with regard to higher temperature of their operation and reduction of the level of air pollution.

Having a calorific value of ca. 30.2 MJ/kg, which is about equivalent to that of coal, PET is readily suited for the incineration process. However, like other plastics its combustion requires 3–5 times more oxygen than for conventional incineration, produces more soot, and develops excessive heat that thus calls for special incineration equipment to cope with these problems.

Several processes have been developed [41–43] to overcome the technological drawbacks of plastics incineration cited above. These include continuous rotary-kiln processes; a process for glass-reinforced PET; a combined system for wood fiber and PET to provide steam to power equipment; and a fluidized system for pyrolysis, in combination with silver recovery from photographic film. Incineration of photographic film raises the additional problem of the formation of toxic halogenated compounds due to the presence of silver halides.

Incineration of PET is usually carried out at temperatures around 700°C, since at lower temperatures waxy side products are formed, leading to clogging, while at higher temperatures the amount of the desirable fraction of mononuclear aromatics in the condensate decreases. A representative sample pyrolyzed under optimum conditions yields, in addition to carbon and water, aromatics like benzene and toluene, and a variety of carbon–hydrogen and carbon–oxygen gases. Studies have been made [44] relating to the formation of dioxines and residual ashes containing heavy metals and other stabilizers. While most problems arising during incineration of PET can be resolved, it is evident that quite a few hurdles remain to be overcome before an economically feasible and ecologically acceptable industrial technical process becomes available.

In conclusion, it may be said that there exists a clear hierarchy in PET-film recycling technologies. Two most important criteria of classification are the degree of purity of PET scrap to be handled and the economics of the process. While for the cleanest PET grade the most economical process is direct reuse in extrusion, for less-clean PET samples it is still possible to reuse them after the modification step (partial degradation, e.g., by glycolysis) at a reasonably low price. More-contaminated PET waste must be degraded into the starting monomers, which can be separated and repolymerized afterwards, of course, at a higher cost. For this operation, mostly the methanolysis process has been exploited industrially. Finally, the most heavily contaminated PET wastes have to be incinerated or brought to a landfill.

2.4 Recycling of Polyurethanes

Polyurethanes are by far the most versatile group of polymers, because the products range from soft thermoplastic elastomers to hard thermoset rigid forms (see Chapter 1 of *Industrial Polymers, Specialty Polymers, and Their Applications*). Although polyurethane rubbers are specialty products, polyurethane foams are well known and widely used materials. While the use of plastics in automobile has increased steadily over the years, a major part of these plastics is polyurethane (PU), which is used for car upholstery; front, rear, and side coverings; as also for spoiler. In fact, about half of the weight of plastics in modern cars is accounted for by PU foams. Accordingly, in addition to production scrap, large quantities of used PU articles are now generated from automotive sources. Though most PU plastics are cross-linked polymers, they cannot be regarded as ordinary thermosetting plastics, owing to their chemical structure and physical domain structure. Thus in contrast to typical thermosetting plastics, various methods are available today for recycling PU scrap and used products.

There are basically two methods for recycling polyurethane scrap and used parts, namely, material recycling (primary, secondary, and tertiary recycling) and energy recycling (quaternary recycling). The former methods are preferred since in this way material resources are replenished. After multiple uses the material can finally be used for energy recovery by high-temperature combustion or gasification.

Among several processes described for PU material recycling, thermopressing and kneader recycling [45] have attracted much attention. By the thermopressing process, granulated PU wastes can be converted into new molded parts, while in the kneader recycling process a thermomechanical operation causes partial chemical breakdown of PU polymer chains that can be subsequently cross-linked by reacting with polyisocyanates. Hydrolysis and glycolysis are important tertiary recycling processes for PU wastes.

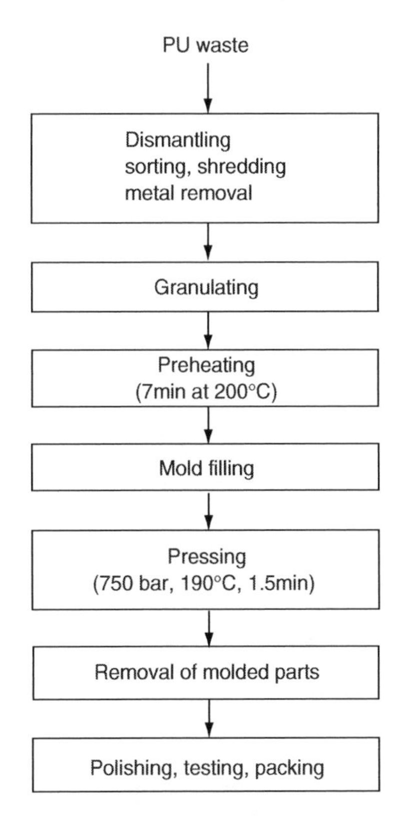

FIGURE 2.5 Reprocessing of polyurethane waste by thermopressing. (After Müller, P. and Reiss, R., *Die Makromol. Chem., Macromol. Symp.*, 57, 175, 1992. With permission.)

2.4.1 Thermopressing Process

Thermopressing, or molding by heat and compression, is a direct method of material recycling that is designed such that elastomeric, cross-linked polyurethanes can be recycled in much the same way as thermoplastic materials [46]. The principle of thermopressing is based on the realization that polyurethane and polyurea granules are capable of flowing into each other and building up new bonding forces under the influence of high temperature (185–195°C), high pressure (300–800 bar), and strong shearing forces. The granules generally used for this purpose have a diameter of 0.5–3 mm. They completely fill the cavities of a mold meaning that moldings with new geometries can also be manufactured.

Unlike injection molding of thermoplastics for which a cold mold is used, in the thermopressing process, the mold is kept constantly hot at a temperature of 190 ± 5°C and no release agent is used for demolding. This relatively simple technique will permit 100% recycling of polyurethane RIM and RRIM moldings, particularly when the formulations of RIM systems to be used in future have been optimized for recycling. The steps in the thermopressing process are shown in Figure 2.5.

The molded parts obtained by thermopressing of granulated PU waste exhibit only slight reduction in hardness and impact strength but significant reduction in elongation at break. The last named property, for example, drops to about 10% of the original value if painted PU wastes are used. Moreover, because of the use of granulated feed, the resulting molded parts lack surface smoothness and thus should be used preferably in those areas where they are not visible. In a passenger car, there are many such parts that are not subjected to tensile stress but require dimensional and heat stability—properties fulfilled by PU recycled products. Examples of application are wheelboxes, reserve wheel covers and similar other covers, mudguard linings, glove boxes, and casings.

2.4.2 Kneader Process

The basic of the kneader recycling process is a thermomechanical degradation of polymer chains to smaller-size segments. The hard elastic PU is thereby converted into a soft, plastic (unmolten) state, which is achieved with a kneader temperature of 150°C and additional frictional heating. This leads to temperatures above 200°C and causes thermal decomposition into a product that is soft at 150–200°C but becomes brittle at room temperature, enabling it to be crushed to powder in a cold kneader or roller press. The resulting powder can be easily mixed with a powder form polyisocyanate (e.g., Desmodur TT or 44 of Bayer) and molded into desired shapes by compression molding at 150°C and 200 bar pressure. The scheme of the recycling process is shown in Figure 2.6.

Partial breakdown of PU network in the kneader results in highly branched molecules with many functional groups necessitating addition of polyisocyanate in relatively high concentration for

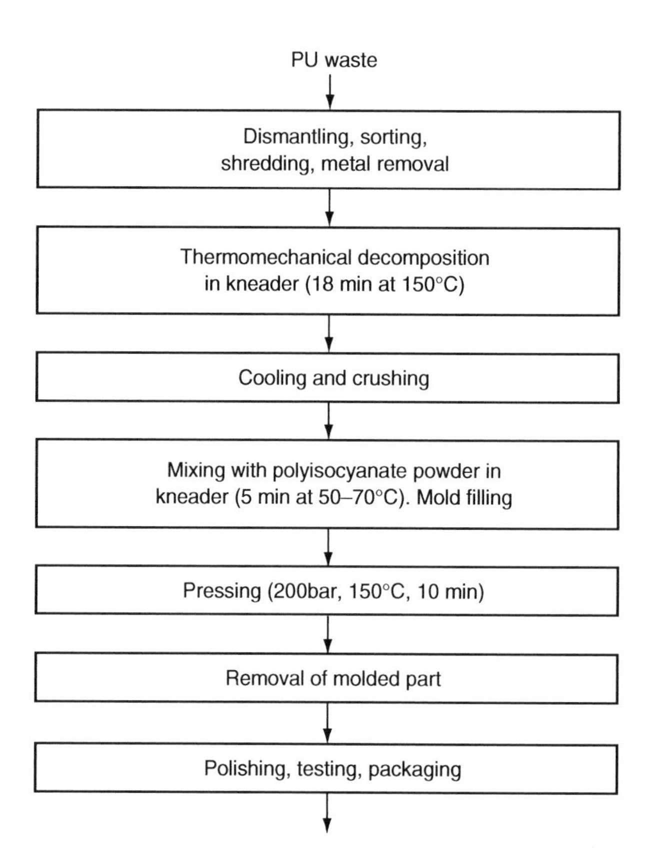

FIGURE 2.6 Recycling of polyurethane waste via partial decomposition in kneader. (After Müller, P. and Reiss, R., *Die Makromol. Chem., Macromol. Symp.*, 57, 175, 1992. With permission.)

subsequent cross-linking to produce molded articles. The process thus yields products of high hardness (with Shore up to 80) and high tensile strength (30 MPa), but small elongation at break (6–8%).

2.4.3 Hydrolysis

Hydrolysis of PU waste results in the formation of polyethers and polyamines that can be used as starting materials for producing foam. In this process, powdered PU waste is reacted with superheated steam at 160–190°C and the polymer gets converted in about 15 min to a liquid heavier than water. The liquid is a mixture of toluene diamine and propylene oxide (polyether diol), the former accounting for 65–85% of the theoretical yield:

$$\sim\!\sim R{-}NH{-}\overset{\overset{O}{\|}}{C}{-}OR'{-}\sim\!\sim \xrightarrow{+H_2O} \sim\!\sim RNH_2 + CO_2 + \sim\!\sim R'OH \tag{2.4}$$

The recovered polyether can be used in formulations for making PU foam, preferably in admixture with virgin polyether [47].

A continuous hydrolysis reactor utilizing a twin-screw extruder has been designed [47] that can be heated to a temperature of 300°C and has a provision for injection of water into the extruder at a point where the scrap is almost in the pulp state. Polyurethane scrap in powder form is fed into the extruder and residence time is adjusted to 5–30 min. Separation of the two components, polyether and diamine, in the product may be effected by fractional distillation, by extraction with a suitable solvent, or by chemical means. The PU foams made from these recycled products can be used in several applications, one example being protection boards for construction sites. Hydrolytic recycling has not, however, found much application, since virgin raw materials are cheaper than the regenerated products.

2.4.3.1 Glycolysis

Extensive studies have been made on glycolytic degradation of PU wastes. In a glycolytic process, powdered PU waste is suspended in a short-chain glycol and hated to a temperature of 185–210°C in nitrogen atmosphere. The glycolysis reaction takes place by way of transesterification of carbonate groups in PU (Figure 2.7). The reaction product is predominantly a mixture of glycols and does not need any further separation of the components, unlike in the hydrolytic process. The cost of producing such recycled polyol is reported to be low enough to make the process economically viable [47].

The mixed polyols resulting from glycolytic degradation of PU waste is suitable mainly for the production of hard foam, such as insulating foam for houses.

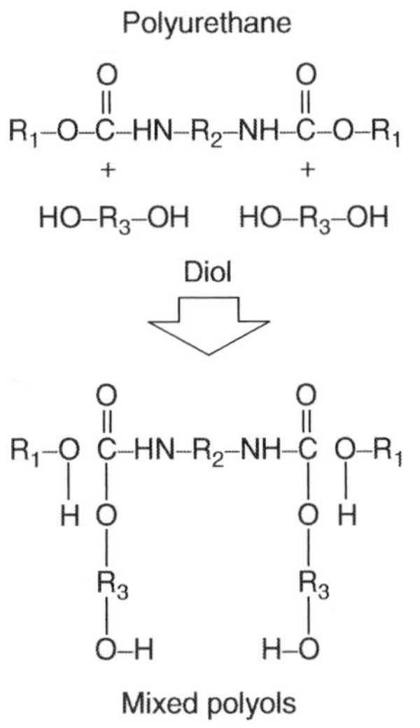

Polyurethane

$$R_1{-}O{-}\overset{\overset{O}{\|}}{C}{-}HN{-}R_2{-}NH{-}\overset{\overset{O}{\|}}{C}{-}O{-}R_1$$

$$+ \qquad\qquad +$$

$$HO{-}R_3{-}OH \qquad HO{-}R_3{-}OH$$

Diol

$$R_1{-}OH \quad HO{-}R_3{-}O{-}\overset{\overset{O}{\|}}{C}{-}HN{-}R_2{-}NH{-}\overset{\overset{O}{\|}}{C}{-}O{-}R_3{-}OH \quad HO{-}R_1$$

Mixed polyols

FIGURE 2.7 Alcoholysis of polyurethane (PU) waste. By the action of small-chain alcohols (e.g., diol), PU is decomposed yielding homogeneous, liquid, and mixed polyols.

2.4.3.2 Ammonolysis

Chemical recycling of polyurethanes by ammonolytic cleavage of urethane and urea bonds under supercritical conditions has been described [48]. It is well known that a number of low-boiling materials give enhance solubility and reactivity under supercritical conditions. Ammonia has a critical point at 132.45°C and 112.8 bar (11.28 MPa) with a density of 0.235 g/cm^3. Being able to act as hydrogen-bond donor and acceptor, it provides good solubility for polyurethanes and

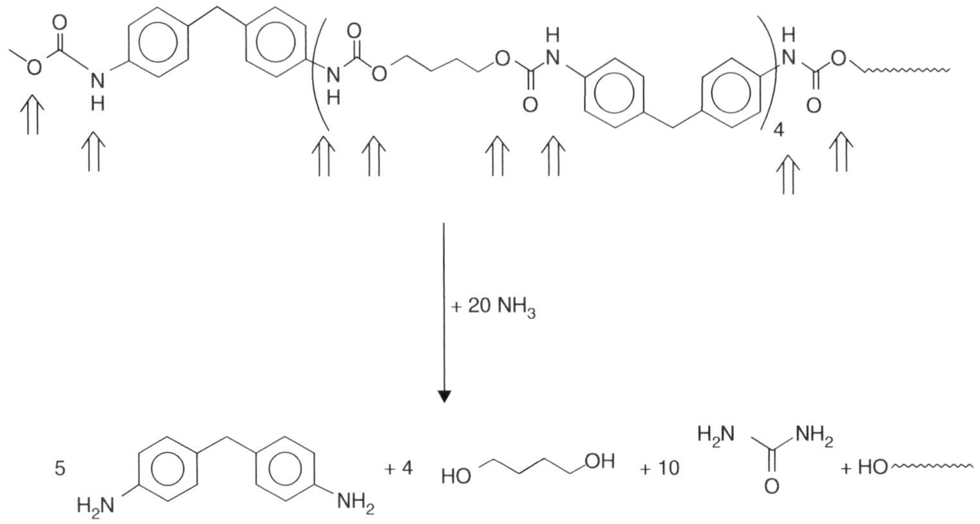

FIGURE 2.8 Stoichiometry of ammonolysis reaction of a polyetherurethane. (After Lentz, H. and Mormann, W. 1992. *Die Makromol. Chem., Macromol. Symp.*, 57, 305.)

dissolves their hard segment domains thus enabling a homogeneous reaction. Ammonia is also a reagent having greater nucleophilicity than, for example, water or glycol is; since it is added in a huge molar excess compared to the urethane or urea groups of the materials to be cleaved, the equilibrium is shifted towards the ammonolysis products. The stoichiometry of ammonolysis reaction of a polyetherurethane is shown in Figure 2.8.

The typical reaction parameters of an ammonolysis process are temperature of 139°C, pressure of 140 bar, and reaction time of 120 min. The ammonolysis reaction transforms derivatives of carbonic acid

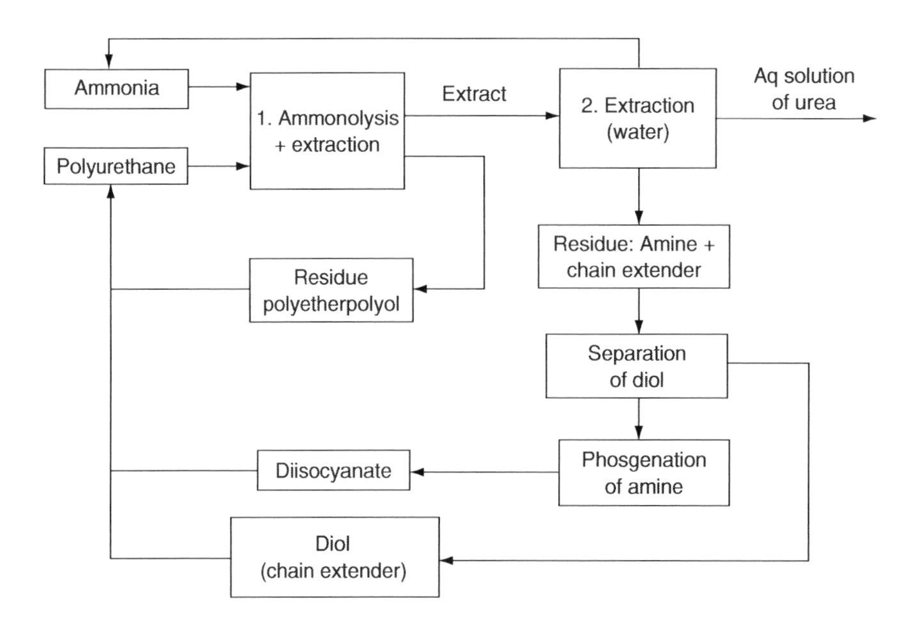

FIGURE 2.9 Flow scheme of a chemical recycling process based on ammonolytic cleavage and separation of polyol by supercritical ammonia. (After Lentz, H. and Mormann, W., *Die Makromol. Chem., Macromol. Symp.*, 57, 305, 1992. With permission.)

into urea. Ether bonds as well as hydroxy groups are inert towards ammonia under the reaction conditions applied. Hydroxy compounds like polyols and diol chain extenders that do not contain ester groups are recovered as such. The C=O fragments of urethane and urea functional groups are converted to unsubstituted urea.

After ammonolysis, ammonia is evaporated and can be reused after liquefaction, while degradation products of polyurethane hard segments (e.g., amines and chain extenders) and urea are removed by extraction. The pure polyol is left in the reactor. It can be removed mechanically or by extraction with liquid ammonia in which it is soluble. The recovered amines can be converted to the corresponding isocyanates and can be reused, along with polyols, in the same applications as before. A flow scheme of the recycling process is shown in Figure 2.9.

Among the various material recycling methods for PU scrap and wastes described above, the thermopressing and kneading processes are especially significant, because these simple processes render the recycling of cross-linked PU products equivalent to that of thermoplastic products. Lack of surface smoothness and some reduction in mechanical properties are to be tolerated, especially when painted PU wastes are recycled. However, good values of E-modulus, structural rigidity, and hot and cold impact resistance permit use of the molded components of recycled PU in many applications, e.g., in unsighted parts of automobiles, instruments, and machineries.

2.5 Recycling of Poly(Vinyl Chloride)

Aside from the polyolefins, poly (vinyl chloride) (PVC) [49] and some other chlorine-containing polymers belong to the most widely applied thermo-plastic materials. There are many applications of rigid and plasticized PVC. In the building sector, for example, very large amounts are used for pipes, profiles for windows, floor coverings, roofing sheets and so on. By the end of the lifetime of these articles, large amounts of scrap have been produced. It is of economic and environmental interest to recycle this PVC waste as much as possible. Disposal of PVC waste by incineration has its special problems. Due to the high chlorine content of PVC, its incineration yields large amounts of HCl gas in addition to the possibility of formation of toxic dioxines and furans. On the other hand, it is a great advantage that many sources produce large amounts of PVC scrap of the same origin and with similar composition, which simplifies the reuse possibilities from a logistic point of view.

Dealing with post-consumer mixed PVC waste involves special considerations. Reprocessing PVC-containing plastics waste without separation will normally entail dealing with mixtures in which large proportions of polyolefins (mainly polyethylene) are present. In view of the poor compatibility of polyolefins with PVC, this is not a particularly attractive practical proposition, with respect to processing and the resulting product. Selective reclamation, i.e., separation from waste mixtures with other plastics, and subsequent reprocessing are complicated by the wide variety of PVC formulations, and the increased susceptibility to heat degradation in reprocessing. The main factors in the latter are the heat history already acquired; the possible presence of polymer already partly degraded in the course of past heat treatments and/or service; and the remaining stability of PVC articles before their recycling, which often necessitates an additional stabilization by addition of heat stabilizers. Moreover, about 1/3 of the used PVC is plasticized by various types of plasticizers. Therefore, for the recycling of such PVC types the concentration of plasticizers should be known. Due to these considerations, it is important to have rather detailed information about a PVC scrap before use.

2.5.1 Characterization of Used PVC

Since several chemical reactions occur during processing and use of PVC, which can change the properties of the polymer, it is necessary to characterize PVC scrap before deciding about the reusability.

Under the influence of heat and light (and also oxygen), PVC chains can be degraded or even cross-linked, which results in changes in the molecular weight and distribution and thus in the mechanical properties of PVC. For determining the molecular weight distribution, gel permeation chromatography is the most applied method, but in many cases the measurement of solution viscosity after separation of all insoluble components, including cross-linked PVC, will suffice.

Because practically no PVC is processed and used without the addition of stabilizers, one should know the residual stability of a used PVC product. For this the best way may be the determination of the hydrogen chloride elimination at 180°C under air or nitrogen [50]. The conversion-time curves so obtained provide indication of the residual stability from the induction period and also enable calculation of the rate of HCl split-off after consumption of the stabilizers. In some cases, however, it may be sufficient to use a simple Congo Red test (e.g., according to DIN 53 418) instead of the apparatus for measuring the HCl elimination.

The dehydrochlorination of PVC results in the formation of polyene sequence that can be responsible for discoloration and also act as starting sites for further degradation and cross-linking reactions. For some applications it may thus be useful to have some knowledge about the unsaturated structures that have been formed in PVC during the use. For this purpose, the investigation of the UV-VIS spectra that give at least semiquantitative information about the dehydrochlorination and the application of ozonolysis [51], which results in cleavage of the unsaturated sequences in PVC, may be useful.

In the case of reuse of plasticized PVC, it is important to determine the residual plasticizer content. This can be obtained by extraction with ether or similar nonsolvents for PVC and determination of the chemical nature of the plasticizers by thin-layer or gas chromatographic methods. The determination of the glass transition temperature by differential thermal analysis also gives information on the efficiency of the residual plasticizers.

2.5.2 In-Line PVC Scrap

Normal recirculation, in the same process, of the clean PVC scrap generated (e.g., edge trim in calendering) is widely practiced, in particular with PVC for noncritical applications. General PVC scrap, both from internal and external sources, is also converted by some processors into such products as cheap garden hose or core composition for cables.

Two possibilities have been investigated for the recycling of die-cutting scrap produced in the processing of PVC sheet: production of secondary sheet and production of extruded profiles or pipes. First the scrap is ground up in a grinding mill. The stabilizer used in the original process is added again in a premixer, and pigments are often added to achieve a uniform, desired color. This premix is fed into the compounding unit.

If the regrind includes rigid, semirigid, and plasticized PVC, the rigid and semirigid fractions can be charged into the compounding unit through a first-inlet opening and plasticated in an initial kneading zone. The plasticized PVC scrap is then fed into this fluxed stock, which ensures the gentlest and most homogeneous processing (Figure 2.10). Any additional plasticizer required is injected directly into the

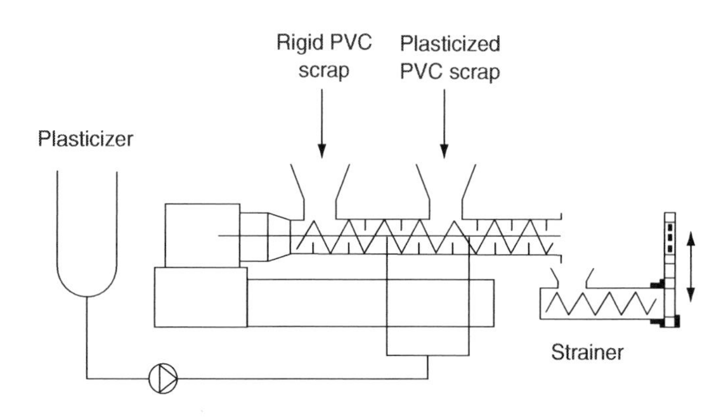

FIGURE 2.10 Recycling of PVC film scrap.

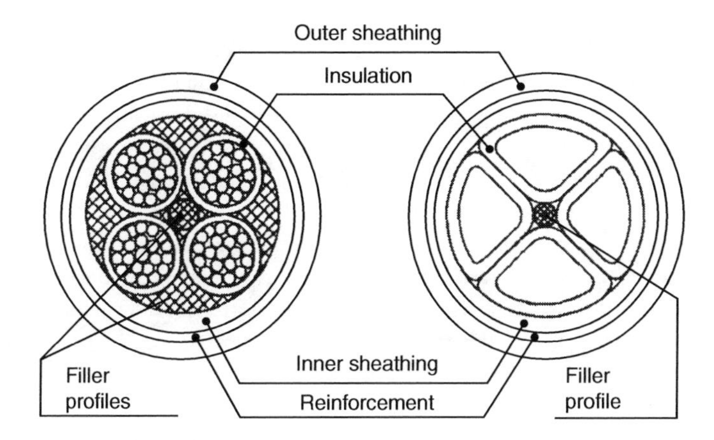

FIGURE 2.11 Schematic of cable design. (After La Mantia, F. P. ed. 1996. *Recycling of PVC and Mixed Plastic Waste*, ChemTec Publishing, Toronto-Scarborough, Canada.)

kneading zone of the compounding unit by a pump. This is done because the plasticizer cannot diffuse in the PVC regrind within a reasonable time, which it can in the case of virgin PVC. It is advisable to use a strainer in order to remove any contamination from the stock. Afterwards, the calendering process is carried out as usual.

For the production of profiles and pipes, the homogeneous stock is pelletized following compounding. The pellets are fed to an extrusion line.

In the cable sector, compounders are often confronted with the problem of recycling copperless insulation and sheathing scrap. An approach that may be taken in this case is to use this scrap for producing filling core mixtures. The purpose of the filling cores is to fill out the cavities between a cable's conductors (Figure 2.11). Since their composition is not subject to any special electrical or mechanical specifications, it is normally made as inexpensive as possible, usually receiving a high level of chalk filler. The PVC in this case acts mainly as binder for the filler. For compounding such cable filler cores, the reground PVC scrap, with a particle size of 5–10 mm, is fed into the first inlet of a compounding unit designed specifically for this application (Figure 2.12). The reground scrap is plasticated homogeneously in the first kneader zone, enabling it to absorb the high filler loading fed into the second inlet opening without any difficulty. For increased flexibility of the filler cores, plasticizer may be injected into the kneading chamber by a pump (see Figure 2.12). The homogenized stock is pelletized following compounding.

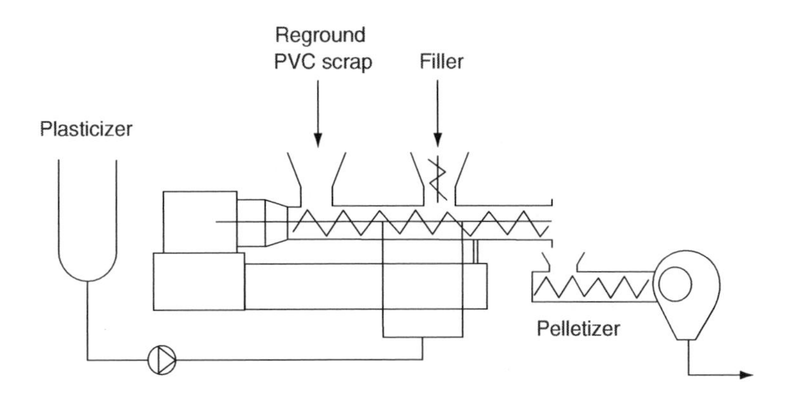

FIGURE 2.12 Compounding of cable filler cores.

2.5.3 PVC Floor Coverings

PVC floor coverings are a combination of a number of constituents that together comprise the recipe for a floor covering. Typical floor covering recipes are for the most part made up as follows: PVC 28–50%, plasticizer 10–20%, stabilizers 0.5–1%, slip agents less than 1%, filler 25–60%, and pigments 1–5%. Further, many floor coverings are provided with additional textile- or glass-fiber—containing carrier layers. Given an average service life of 10–17 years, old PVC coverings represent a large reserve of raw materials.

To exploit these large reserves of raw materials on an industrial scale, one needs a network for systematic collection of old coverings, a system of transport logistics, and the ability to build technically feasible recycling plants. In order to undertake this work in Europe, about 20 producers of PVC raw material and floor coverings from Germany and other European countries joined together in April 1990 to form the Society for the Recycling of PVC Floorings.

The main operations carried out in a recycling plant for old PVC floor coverings are sorting, cleaning, shredding, purifying, powdering, mixing, and packaging [52]. The purification unit essentially comprises a hammer mill and a downstream vibrating screen. The function of the hammer mill is to knock off any residues of screed or adhesive still adhering to the floor covering, and the vibrating screen then separates off these residues. The shredding material that has been purified in this way is first passed to a cutting mill, which enables it to be precomminuted to granules that are first homogenized in a mixing silo and purified by means of zigzag sifters before being processed to PVC floor covering powder in the powder mill. The powder can be upgraded by the addition of plasticizer, PVC, or filler to give powder recipes suitable for calendering to make new floor coverings; this depends on the quality of the batch in question. Other possible uses of the powder are for products, such as mats for cars, mud flaps, and soft profiles.

2.5.4 PVC Roofing Sheets

A major use of PVC in the building industry is for roofing sheets. These sheets are produced on calenders and contain in most cases two plasticized PVC foils that are reinforced by glass fiber or polyester fabrics. The used sheets show different properties, depending on whether they are applied under direct influence of light and weather conditions and on the fact that in some cases these sheets on the roof are loaded with gravel or coarse sand.

After a lifetime of 10–20 years, the roofing sheets have to be replaced by new ones, which is normally done by the same firm(s). Therefore, the recycling of used roofing sheets is rather easy from the logistic point of view and has been common practice for many years. New PVC roofing sheets may contain up to 10% of the recycled material [52].

2.5.5 Post-Consumer PVC

Some post-consumer PVC sources are water, food, pharmaceutical, and cosmetic bottles, and film. Another significant source of post-consumer PVC is used electric cable, coming principally from plant demolition and, to a lesser degree, from manufacturing scrap and offcuts.

Most end-use markets for recycled plastic bottles require that they be separated by resin type and color. This ensures high end-use value for new products incorporating substantial amounts of the recycled resin. PVC bottles, like PET bottles, are very recyclable. Manual sorting of nonpigmented PVC and clear PET bottles is difficult because they look alike. When the two types are received commingled, the reprocessor can experience quality deficiencies due to rheological incompatibilities between these two resins. Therefore, all attempts to separate and remove these two resins must be made prior to recycling.

Manual sorting techniques are inadequate to meet the market's needed quality standards, so new techniques have been engineered that will detect and separate bottles made from either of these two resins. A simple device senses the presence of chlorine atoms as a means to detect PVC bottles.

Once detected, PVC bottles are pneumatically jettisoned from the commingled bottle feed-stream by a microprocessor-based air-blast system.

The step after sorting is baling or granulation. Granulation is the preferred method of intermediate processing since the material so processed commands the highest market value. For upgradation of the resin from the recycled PVC bottles, several steps are explored depending on the results of characterization tests as discussed earlier. These include incorporation of virgin resin (10–90%), restabilization against UV and heat, and incorporation of processing aids, impact modifiers, lubricants, plasticizers, and antioxidants.

The recovery of electric cable is long established because of its valuable copper content. After this conductor material has been extracted, the residue consists of sheathing and insulation that may contain rubber and polyethylene as well as PVC. These other materials can be largely removed from grinding, by flotation, vibration, and filtration, but rubber is especially difficult to remove entirely, so that applications for material recycled from cables containing it are limited to areas such as car mats and carpet underlay.

2.6 Recycling of Cured Epoxies

Thermosetting plastics are difficult to dispose because of their network structure. Chemical recycling is a promising route for converting these plastic wastes by returning them back to their original constituents. However, thermosetting resins are usually reinforced by reinforcement such as glass fiber to modify their brittleness and increase their strength, forming composite materials with complex structure. The presence of reinforcement in the cured composite thus makes the recycling of the matrix resin more difficult.

An approach to chemical recycling of amine cured epoxies using nitric acid solution has been proposed [53,54]. In order to investigate the practical applicability of the proposed research, glass fiber-reinforced bisphenol F type epoxy resin (cured with 4,4′-diaminodiphenylmethane) was decomposed in nitric acid solution and the decomposed organic products as well as the fiber were recovered. In a typical experiment, the glass-reinforced epoxy composite was cut into small pieces and kept immersed in 4 M nitric acid at 80°C till the matrix resin dissolved completely, yielding a yellow solution and leaving behind the inorganic (glass) residue which was separated and recovered. When the yellow solution was cooled in ice no crystal was formed. However, if nitric acid solution of higher concentration, such as 6 M, was used for immersion, crystals separated out because of breakage of the main chain of epoxy resin and subsequent nitration under the attack of nitric acid [55].

The yellow solution was subjected to neutralization with sodium carbonate, extraction, refinement, and drying to obtain neutralized extract (NE) whic was then repolymerized to prepare the recycled resin. Since NE could contribute hydroxyl groups to bond with phthalic anhydride (curing agent), it was employed to substitute a part of epoxy resin. The proportions of NE addition ranged from 5 to 30 wt% (the ratio of weight of NE to the total weight of NE and epoxy resin). The process of neutralization and refinement of the acid extract is presented in Figure 2.13.

2.7 Recycling of Mixed Plastics Waste

Commingled plastics currently represent an estimated two-thirds of today's recycled plastics streams. That fraction can be expected to shrink somewhat with the development of more successful identification and segregation technologies in the future. However, commingled plastics streams will continue to make up a significant volume for several reasons: proliferation of grades and types of commodity; profusion of polymer blends and alloys; contamination of recycle plastic parts with metals, coatings, and laminates; and practical cost considerations. Mixed plastics wastes can be divided into two groups depending on their source: mixed plastics from household or municipal solid wastes and plastics from industrial sectors.

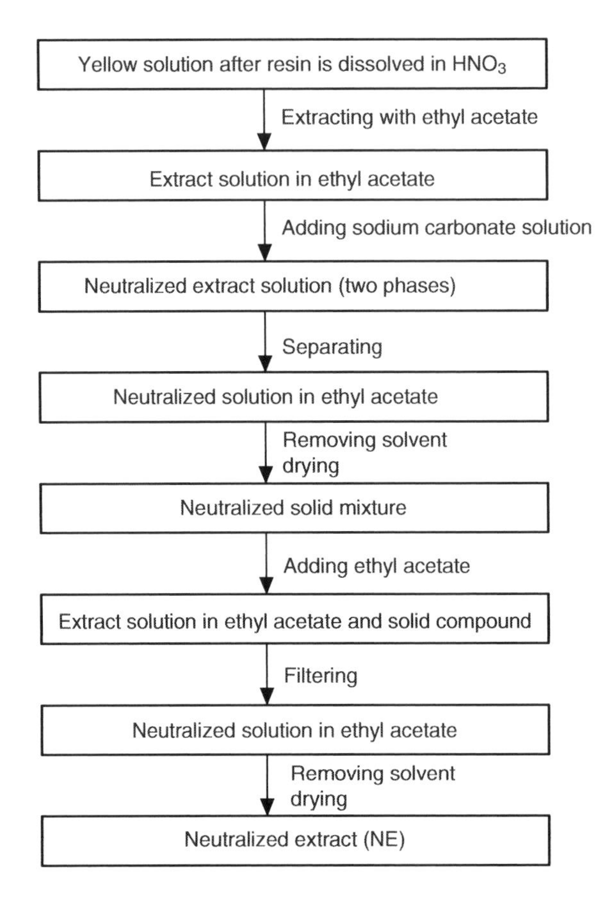

FIGURE 2.13 Process of extraction of epoxy resin dissolved in nitric acid and neutralization of the extract. (After Dang, W., Kubouchi, M., Sembokuya, H., and Tsuda, K., *Polymer*, 46, 1905, 2005. With permission.)

The first category (post-consumer mixed plastics) involves the articles that are used in food, pharmaceutical and detergent packaging, shopping, and others. The majority of these are composed of films, sheeting, strapping, thermoformed trays, as well as a variety of bottles for soft drinks, food, and cosmetics. There are mainly five different polymers—PE, PP, PS, PVC, and PET—that contribute to the total amount of plastics waste. The composition of mixed plastics can change depending on the regional habits and the seasons of the year. Also the mode of waste collection can influence its final composition.

The category of postindustrial wastes concerns articles like the products of the car, furniture, and appliances industries. The problem of these sectors is a wide variety of engineering materials and a high number of components employed to build a final system.

2.7.1 Direct Reuse

A direct solution to the problem of plastics disposal can be the reuse of a heterogeneous mixture of plastics directly obtained from an urban collection. Today there are extruders specifically designed for reprocessing post-consumer and postindustrial waste materials. The waste material can have many forms and can range in bulk density from approximately 1 to 35 lb/cu ft. The form, bulk density, moisture content, contamination level, and process-temperature restrictions all affect the design of the extruder to be used. For example, due to the presence of PVC resin, the melting temperature must be kept below 210°C and the barrel residence time must not exceed 6 min. Furthermore, in the mixture, the relatively high content of semicrystalline polymers like PET, whose melting point is above the processing temperature, influences the extruder design or the final properties of the manufactured product. Large injection gates and mold channels must be used in order to avoid undesirable occlusions in the channels.

Standard single-screw extruders are no longer adequate to recycle or reclaim this wide range of materials in a cost-effective manner. There are now special extruders designed to process the lighter-bulk-density materials. Low-bulk-density materials are the various forms of film, fibers, and foams commonly used in the packaging industry. Due to their low-bulk density, such materials typically require an auxiliary device to facilitate proper feeding into the extruder throat. There are several varieties of such feeding mechanisms available [56]. Two of them are a rotating screw-type crammer and a piston-type ram. The crammer and ram systems both act on the same principle; that is, an auxiliary feeding device is used to convey and to compact the low-bulk-density materials into the feed section of the extruder screw.

The screw-crammer system uses a conical hopper with a screw that is driven by a separate gear reducer and variable-speed drive motor. The output and effectiveness of the crammer are determined by the screw configuration and the available speed. The ram-type system, on the other hand, uses a pneumatic ram to stuff material into the screw. The ram is a piston-driven unit with the stroke timing adjustable by setting a series of timers located in the control panel. The feed section used by the ram system has an opening that is 12–14 times larger than that of a standard screw extruder. This allows low-bulk-density material to flow freely into the feed throat where the ram can compress it into the screw. Depending on the extruder size, the ram can compact materials with a force of 2000–9000 psi.

Feed materials usually need to be supplied to either the crammer or the ram system in a chopped form. The size and bulk density of the chopped particles affect the performance of the crammer and ram, and thus ultimately the output of the extruder. Both these systems can also be used to process higher-bulk-density products.

A third method of processing low-bulk-density materials is through the use of a dual-diameter extruder [56]. This system has two distinct sections: a large diameter feed and a small diameter processing section. The large-diameter section acts as a cramming device—compacting, compressing, and conveying the feed material—while the smaller-diameter section is used to melt, devolatilize, and pump the extrudate into a die. In the feed section, the screw can have deep flights, allowing low-bulk-density materials to flow freely, while in the processing section the screw resembles that of a typical extruder. The screw is available in one of several configuration: single-stage, two-stage, or barrier design. Depending on process needs, these designs optimize output and raise product quality.

The feed section of a dual-diameter extruder can be equipped with feed-assist components that in some cases work in conjunction with specially designed screws to allow processing of a wide variety of feedstocks that are fed to the machines in roll form. Among these possible feedstocks are loose bags, handle cutouts from bag making operations, and continuous web products such as blown and cast film scrap. This ability eliminates the cost of shredding, grinding, and densification of many materials. A dual-diameter extruder is also capable of processing materials with a high-bulk density. The crammer, ram, and dual-diameter systems do not differ widely in equipment costs, or production rates.

It is well established that a strong incompatibility is typical of polymers usually found in commingled waste (PE, PP, PET, PVC, and PS). This incompatibility gives rise to materials that have inferior mechanical properties, particularly with regard to tensile, flexural, and impact strengths. This means a strong limitation of applications, in particular in the case of thin walls and manufactured products that have to work under flexural and tensile stresses. However, by adding to the mixture specific components like other polymers from homogeneous recycling, fillers (talc), fibers, or promoters (compatibilizers) that increase the compatibility, it is possible to improve the tenacity or stiffness, product aesthetics, and processability.

Addition of glass fibers, for example, is found [57] to yield products with very high stiffness (e.g., elastic modulus $E \simeq 2800$ MPa with 30% glass fiber), higher than that with talc ($E \simeq 1250$ MPa with 20% talc) and far better than that of the original mixture ($E \simeq 950$ MPa). Addition of LDPE and styrene-butadiene-styrene copolymer, on the other hand, improves the tenacity (showing, typically, a 30–90%

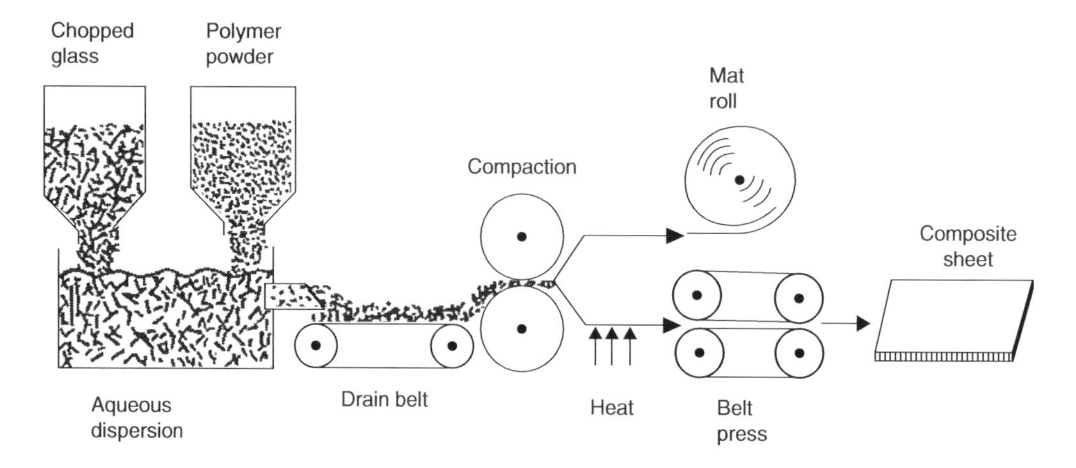

FIGURE 2.14 Schematic of Radlite technology.

increase in elongation at break). Extruded profiles have thus been made that can be employed to build benches, garden tables, bicycle racks, fences, and playing facilities for park. Coextrusion technology can be used very effectively to improve surface properties like puncture, impact and weather resistance, as well as appearance. One interesting application for mixed plastics, because of their large market volume, is the production of injected tiles for paving [57].

A new and exciting technology has been developed in the fabrication of composite materials made with commingled plastics. It followed the discovery in 1986 by the scientists and technologists at GE Plastics that useful products could be fabricated using Radlite technology (Figure 2.14) if the powdered feed consisted of two or more resins. While compatibilization of dissimilar resins is typically brought about by chemical means, this is not the case with Radlite technology products where compatibilization appears to take place by physical means, i.e., the binding of dissimilar resin domains through the fibers. A conceptual model of physical compatibilization is shown in Figure 2.15.

The rolls or sheets made by Radlite technology using commingled plastics and chopped glass fibers (0.25–1.0 in.) can be converted to finished parts by conventional forming technologies such as compression molding (typically 200–290°C, 3–5 MPa, 2–4 min). These are fully consolidated, essentially void-free products with specific gravities that would be calculated on the basis of resin type and glass fiber content. The physical properties of the consolidated structures from commingled plastics are in general similar to standard grades of SMC composites.

Foamed network structures (lofted structures) are also prepared in a compression press using Radlite technology. In this case, after applying full temperature and pressure for the requisite 2–4 min, the platen gap is opened 1.5 times the original setting (for primarily open-cell structure) or 1.1–1.2 times the

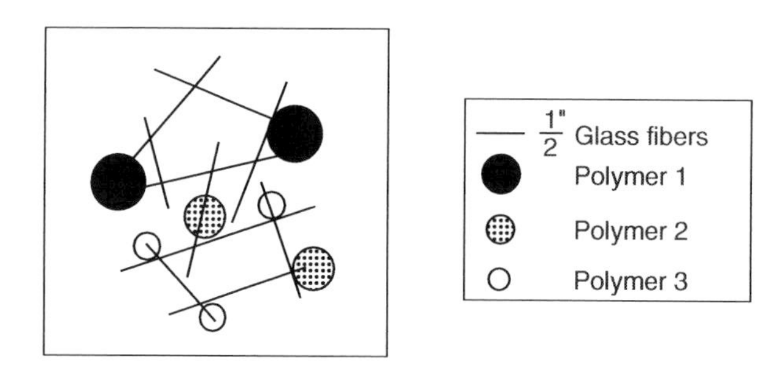

FIGURE 2.15 Conceptual model of physical compatibilization.

original setting (for primarily closed-cell structure). Foamed structures have the aesthetic and physical appearance of particle board, a common material of construction. Application possibilities of products made by the Radlite technology include highway signs and sound barriers, substructures for bathtubs and shower stalls, pallets and tote boxes, and flooring products and building fascias.

A novel application of mixed plastics is to toughen road surfaces. In 1986, the Ragusa Laboratories of ECP Enichem Polimeri in Italy investigated the use of mixed plastics waste to reinforce bitumen. Sorted municipal waste with a polyethylene content of approximately 60% was mixed with bitumen in varying proportions up to 20%. The properties of the resultant bituminous concrete were improved in two important ways: better wear resistance and raised softening point. Use of bitumen modified with mixed plastics waste of high polyethylene content as an experimental road surface under heavy traffic has established its notable superiority over unmodified bitumen for road surfacing.

2.7.2 Homogeneous Fractions

A widespread solution, in terms of application and market volume, could be the recycling of single materials or homogeneous fractions obtained from a separation process of the mixture. In fact, the samples obtained from single homogeneous fractions show a general performance far greater than that of samples produced from mixed plastics. Separation of post-consumer mixed plastics (municipal waste) into four fractions—polyolefins (PO), PS, PVC, and PET—is commonly adequate.

The improvement of tenacity is, in particular, evident when considering impact resistance of PO as compared to mixed plastics. Samples subjected to impact tests show an increase in elongation at the breaking point from 7% to above 100% [58]. The samples of recycled PVC fraction are comparable with those of a common virgin, with only marginal reduction in mechanical properties.

With regard to PET fraction, the potential applications are strongly dependent on its purity (as discussed earlier). Applications like films, fibers, or straps are not recommended when a high concentration of impurities are present. In this case, the PET fraction can be employed for structural applications as an engineering polymer with the addition of other components like glass fibers, impact modifiers, and/or nucleating systems. However, reuse of the PET fraction implies that the amount of residual PVC must be kept below 50 ppm to avoid undesirable polymer degradation that results in poor surface appearance and loss of mechanical properties of the manufactured products.

Mixed waste consisting of PET and PE can be converted into useful products using compatibilizers such as LDPE and LLDPE with acid and anhydride groups grafted on the backbone. A range of products can be made with such compatibilized blend, e.g., office partitions, roofing, slates for benches and chairs, and generally any extruded or molded sections needing mechanical load-bearing capacity similar to aforesaid applications.

With advances in cleaning, sorting, and other recycling technologies, more products with recycled plastics content are being manufactured. Some recent developments include using recycled PP and HDPE to produce a wide range of products. For example, multimaterial PP bottle scrap (typically, 90% PP, 5% ethylene-vinyl alcohol barrier resin, and 5% olefin adhesive) can be added in varying amount (3–12%) to recycled HDPE and processed on a single-screw extruder to form pellets for compression molding to a range of products. The multimaterial resin can also be sandwiched between two layers of virgin HDPE using a three-layer extrusion blow-molding process. Bottles made in this way from 75% virgin resin and 25% post-consumer blend are found to be suitable for normal commercial trade [59].

2.7.3 Liquefaction of Mixed Plastics

There have been many research activities on plastics liquefaction because oil is easy to store, transport, and use. Most promising among them is the liquefaction technology jointly developed by the Japanese Government Industrial Development Laboratory (Hokkaido), Mobil Oil Corporation, and Fuji Recycle [60]. The process features a combination of thermal and catalytic cracking using a proprietary Mobil

ZSM-5 catalyst. It can treat polyolefinic plastics, PE, PP, PS, or their mixtures, producing relatively low pour and highly aromatic liquid at a yield of about 85%. The produced oil contains many aromatics including benzene, toluene, and xylene.

Waste plastics are crushed, washed, and separated from other plastics that cannot be liquefied (e.g., PVC) by utilizing the difference of specific gravities against water. Plastics that can be liquefied mostly float in water while plastics that contain a lot of chlorine, carbon, and oxygen have high specific gravities and sink in water. However, some PVC floats and is recovered with PE or PP. Therefore, after separating, the feedstock for liquefaction may still contain 3–7% PVC. Fuji Recycle has developed liquefaction technology to treat such PVC contaminated mixtures [60].

For liquefaction, polyolefinic plastics are warmed to about 250°C, melted and transferred to the melting vessel by a heated extruder. In the melting vessel, plastics are further heated to about 300°C by heat transfer oil and transferred to the thermal cracking vessel. In the thermal cracking vessel, melted plastics are hated to about 400°C by the cracking furnace. The thermally cracked gas phase hydrocarbon passes through the catalytic reactor containing ZSM-5, where it is cracked and converted to higher-value hydrocarbon. The recovered liquid and gas are separated by cooling and the gas is used as in-house fuel.

Because of the pore structure of ZSM-5, the produced hydrocarbons are composed of low molecular species (4 carbons to about 20) which are in the gasoline, kerosene, and gas oil boiling range. In comparison, the carbon numbers of hydrocarbons produced only by thermal cracking range from 4 to 44. Polystyrene in the feedstock enhances the yield of ethylbenzene, toluene, and benzene, while producing gas that is predominantly propane/propylene.

2.8 Post-Consumer Polyethylene Films

Driven by consumer and legislative pressures, post-consumer film recycling has gained momentum and is now one of the fastest-growing segments of the recycling industry. post-consumer films consisting of LDPE, LLDPE, and HDPE, which are accounted for mostly by grocery sacks, stretch and shrink wrap, agricultural film, packaging, and blow-molding drums, and have thicknesses ranging from 0.2 to 5.0 mils. Film recycling was first developed in Europe for agricultural film, which is relatively easy to process due to its high-bulk density and minimal contamination. Today, interest in recycling systems for film of varying thickness and resin types has changed the design of a recycling line to one that uses both high and low-bulk-density material.

The primary challenge to recycling film is contamination. According to an industry estimate, up to 25% of all grocery bags are contaminated and require very thorough washing in the recycling process. A typical route for recycling plastic bags is shown in Figure 2.16.

Bales weighing 600–800 lb, collected by commercial haulers, film-generating business, or the solid-wastes-handling industry, are first fed into a breaker-shredder. The shredded material then passes over a vibratory conveyor or through metal-detection systems to remove ferrous and nonferrous contaminants. Film is next sent to a sedimentation tank for removal of rocks or dirt before going into the granulator

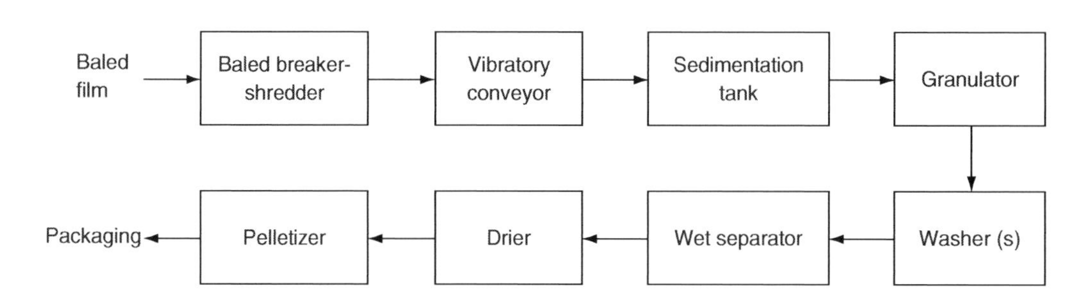

FIGURE 2.16 Typical route for recycling plastic bags.

(usually wet) for size reduction. Granulated into flake, the material then passes to a hot wash for removal of glue, residues, and remaining labels. Depending on the feedstock and the extent of contamination, the film may go through more than one wash cycle.

After the wash, the material moves to the separation stage. Two most common methods for separating plastics by resin type are float-sink and hydrocyclone. The float-sink method is less costly but requires a relatively large tank size in order to realize volume effectiveness. A hydrocyclone is more costly, but uses less water, has no moving parts, and generally takes up less space than the older float-sink method.

The flaked material from the wet separator, described above, goes through a drying and dewatering stage to an extruder, where in the melt phase the plastics can be mixed with dyes and other product enhancements, and then filtered and forced through a die for fabrication of free-flowing pellets. Depending on the mixture of LLDPE, LDPE, and HDPE, the pellet can be tailored for specific markets. Low-quality pellets are often sold as a commodity in the general marketplace. High-quality blends of polyethylene are suitable for many nonfood film applications.

One major market for recycling film is packaging. Many consumer-products companies have already turned to detergent bottles and other packaging made with significant percentages of recycled HDPE. Another fast-growing outlet is coextruded blown film used in trash bags, in which post-consumer resin is sandwiched between virgin layers of high-molecular-weight HDPE. The coextrusion technology allows high percentages of reprocessed material to be incorporated into virgin resin.

In many countries, legislation is a key driving force behind greater recycling efforts. Several states in the U.S. have passed recycling content standards mandating that virgin plastics used in some applications, such as grocery and trash bags, contain a certain percentage of recycled material. In California, for example, trash bags are required to have 30% recycled post-consumer content. Germany requires that 64% of all packaging materials be recycled. Under the German system, all types of plastics packaging are collected together and subsequently segregated into several categories: rigid containers; films; cups, trays, and blister packaging; and foamed material. The materials are then offered back to industry for recycling purposes at no charge.

2.9 Recycling of Ground Rubber Tires

Discarded tires represent a significant component of the overall plastics recycling challenge. They are an easily segregated, large volume part of the waste stream and present their own, somewhat unique, waste recycling problems. Some of the methods of utilizing scrap tires that have been investigated [61–63] are: burning, pyrolysis, use in cleaning up oil spills, road surfaces, roofing materials, and playground surfaces. While some of these approaches have been put into practice, the scrap tire disposal problem is clearly a case where supply far exceeds available use, pointing to the need for new methods of utilization and/or technological advances to extend the existing ones.

One area that has the potential to utilize large volumes of discarded tires is the need for a filler in polymer composites. Although the use of ground rubber tire (GRT) as a filler in polymer blends is a potentially attractive approach, it is fraught with a number of difficulties. Generally, when the large GRT particles are added to either thermoplastic or thermoset matrices, there is a large drop in mechanical properties, even at relatively low filler loadings [64]. Since the approach here is to use the GRT as a low-cost additive, and as there are a number of other materials competing in this regard, overcoming this large drop in properties has to be accomplished with little added cost (both in terms of additives and additional processing). This has proven to be quite a challenging task.

In order to be used as a filler in polymer composites, tires are first ground into a fine powder on the order of 100–400 μm, which is accomplished typically through either cryogenic or ambient grinding. The large rubber particle size used in GRT composites is reported to be one of the two major factors (the other being adhesion) contributing to the poor mechanical properties generally observed for GRT-polymer composites. In general, a low particle size is desired for optimum composite properties.

In GRT-polymer composites, however, the particle size is quite large. Since there is very little breakdown of the particles under normal melt blending conditions due to the highly cross-linked nature of GRT, the particle size is to be controlled only by the grinding process, which in turn, is influenced by process choice and economics.

In order for GRT to be used as an economical filler, the particle size has to be kept as large as possible to minimize grinding costs. Typically, the lower limit on particle size necessary to produce economical composites lies in the 40–80 mesh ($\simeq 400 - 100$ μm) range, while for rubber toughening applications it is generally reported [65] that the optimum particle size for toughening brittle polymers is in the 0.1–5 μm range. Thus the size gap is large. Though it adds to the cost, there may be some advantage in going to smaller particle sizes if significant gains in mechanical properties are realized. The detrimental effects of adding GRT to cured rubbers decreases as the particle size is decreased [66]. For GRT recycled back into tires, for example, the detrimental effects are almost eliminated [66] with the use of ultrafine (20 μm) rubber.

As mentioned above, simple addition of GRT to most polymers results, in general, in significant decreases in mechanical properties due to large particle size and poor adhesion. Although some of these materials may find limited application in low-level usages, there is clearly a need to improve on the properties of GRT-polymer composites for them to become a large-volume material. Since lowering particle size to effect any substantial improvement in material properties adds significantly to grinding costs, strategies for overcoming the deleterious effects of adding GRT to polymers have focused on methods of improving adhesion.

The poor adhesion is, at least in part, due to a high degree of cross-linking in the GRT particles. The highly cross-linked nature of the particles inhibits molecular diffusion across the interface so that there is little or no interpenetration of the phases, resulting in a sharp interface. There have been a number of reports of processes that claim to improve properties of GRT-polymer composites through enhancing adhesion. The use of an aqueous slurry process using a water-soluble initiator system to graft styrene to GRT has been reported [67]. The styrene-grafted GRT particles are found to be give composites with properties superior to straight mechanical blends.

Precoating of GRT particles with ethylene/acrylic acid (EAA) copolymer is found to improve the mechanical property, which is attributed to an interaction (H bonding) between the EAA copolymer and functional groups on the GRT surface, resulting in increased adhesion [68]. Thus a blend of 40 wt% EAA coated GRT particles (4 wt% EAA) with LLDPE was shown to have impact and tensile strengths 90% of those for pure LLDPE, representing increases of 60% and 20%, respectively, over blends with uncoated particles. The use of maleic anhydride grafted PE (PE-*g*-MA) resulted in increases in the impact strength of LLDPE-GRT composites of as much as 43%, without the need for a precoating step [68].

Electronic spectra for chemical analysis (ESCA) of GRT surface reveal an oxygen surface content of 5–15%, which may indicate the presence of –OH or –COOH functionalities. Since the epoxy group readily reacts with a wide range of functional groups such as –OH, –COOH, –SH, $-NH_2$, the use of ethylene-*co*-glycidyl methacrylate (EGM) as a coupling agent [68] has been investigated. A significant increase in impact behavior has been observed. It is seen that judicious selection of a compatibilizing agent can lead to composites with quite reasonable mechanical properties at significant levels of GRT (as high as 50–60 wt%). Because added compatibilizer levels are low (4–7 wt%) and no specialized processing steps are necessary, these higher-value composites can be produced at little additional cost over simple GRT-polymer blends.

It has been observed recently [69] that special treatment of GTR by bitumen confers outstanding mechanical properties on thermoplastic elastomers (TPEs) produced using the treated GTR. Typically the reclamation of GTR by bitumen is carried out by preheating the GTR/bitumen blend (1/1 by weight) at 170°C for 4 h in an oven, followed by rolling on mill rolls at about 60°C for 40 min. Thus, high performance TPEs, based on recycled high-density polyethylene, ethylene-propylene-diene monomer (EPDM) rubber, and GTR treated with bitumen has been prepared. It has been concluded that bitumen acts as an effective devulcanizing agent in the GTR treatment stage. In the subsequent steps of TPE production, bitumen acts simultaneously as a curing agent for the rubber components (EPDM/GTR) and as compatibilizer for the blend components.

2.10 Recycling of Car Batteries

Polypropylene (PP) is obtained in segregated form as casing fragments from reprocessing of used lead-acid batteries from automotive applications. Because the casing makes up about 7% of the total battery and the used batteries are recycled primarily for lead recovery, PP is obtained without additional cost and in substantial quantities to warrant the operation of a plastics recycling plant.

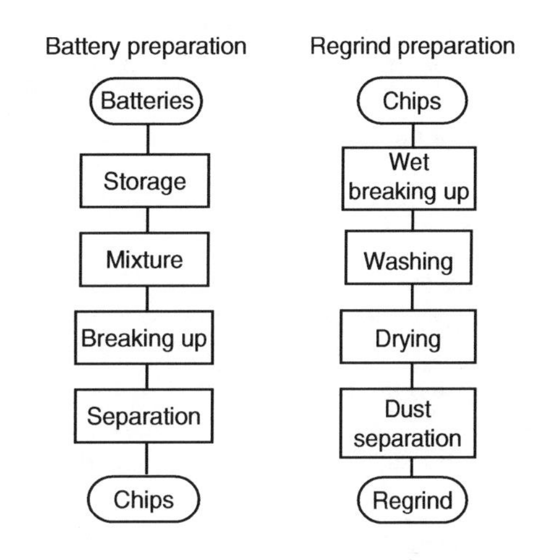

FIGURE 2.17 Process steps in preparation of polypropylene regrind.

In the first step of a typical recycling operation, the batteries are processed through a crushing and separation system operating on the TONOLLI principle (Figure 2.17), which has been successfully employed in various battery recycling plants in Europe and North America. The heavy fraction (lead, lattice metal) and ebonite are then separated from the light fraction (PP and impurities). The PP at this stage has a purity of 97%, which is still insufficient for its further processing. It is therefore sent to an upgrading stage, where it is further reduced in size in a wet-type rotary grinder and subsequently separated from water by sedimentation. After passing through two series-connected driers and a cyclone separator, the PP is available as so-called regrind with a purity of 99.5%. As the regrind consists of various types of PP differing in their formulation, molecular composition, and stabilizer content, it has a broad spectrum of characteristics. Suitable mixing can be done to obtain an intermediate product with a narrowed range of statistically uniform product characteristics.

In the next step, the regrind is routed to a compounding plant where—with controlled addition of additives, polymers, and fillers—the feed mix can be adjusted to suit the specific customer requirements. This feed mix is then metered into a special twin-screw kneader where it melts under the dual action of an external heater and internal shear forces, producing a homogeneous compound. Volatile matter is vented and unmolten impurities are filtered out. The melt is subsequently palletized in a melt granulator and the resulting granulate is quenched in a water bath, centrifuged, and finally passed through a hammer mill to break up lumps. The end product is a granular secondary raw material and suitable for injection molding.

2.11 Plastic Recycling Equipment and Machinery

While the plastics recycling activity, driven by consumer and legislative pressures, is all but certain to increase, the key variables in the rate of growth are the plastics industry's ability to develop an economical material-collection infrastructure and to improve the methods for handling and processing of contaminated scrap. Techniques for selection and recycling of post-consumer plastics are, however, closely related to the characteristics of plastics containers consumption, which vary greatly according to the geographical areas and the relevant law regulations governing activities in this sector.

Consumption features play a major role in the choice of the materials to be recycled. In the United States and Canada, the materials chiefly recycled are PET bottles and PE containers; in France, on the contrary, recycling of PVC got priority on account of the large quantity of such material used in the packaging of drinks. In Australia, recycling includes primarily PET and PE, whereas in Japan it is mainly PET. In Italy, the plastics recycled are mostly PET, PVC, and PE. In short, the material to be recycled and the enforced legislation determine the choice of collection system. In many countries such as United States, Canada, Australia, France, Austria, and Switzerland, some fractions of post-consumer plastics are collected in the most homogeneous way possible. In other countries, plastics are collected more heterogeneously, that is, different types of plastics out of different types of manufactured articles, such as foil, containers, bottles, are collected together.

The outcome of the collection system constitutes the raw material for the recycling process. The degree of purity of this raw material evidently depends on how selective the collection is.

2.11.1 Plastocompactor

For voluminous scraps such as light film, textiles, fleece, and foam, it is advantageous to increase the density, which may be in the rage of 20–40 kg/m^3, to about 400 kg/m^3 for transport reasons and for further processing. With thermoplastics and thermoplastic mixtures, a *plastocompactor* can be used for this purpose. It is, however, more useful for homogeneous fractions of thermoplastics. The process agglomerates the material without plastifying it. By heating the material locally for a short period to a temperature above the softening point, the soft components begin to adhere. The material is then compacted into a condition very similar to the virgin material.

In a typical agglomeration plant, the loose material is usually fed to a granulator, which is also fitted with a nip roll device for feeding continuous material such as fleece or film from the roll. A blower transports the flakes from the granulator to a holding silo. The discharge screw in the silo transports the flakes to the feed hopper from where they are carried by a blower to the plastocompactor. A dosing and pressing screw feeds the actual agglomerator part of the machine. This comprises essentially two discs—one rotating and the other stationary—between which the flakes are compacted by using heat from friction and pressure. The agglomerate leaves the discs through the outer gap in the form of warm soft sausages and is cooled immediately by an air stream. It is then fed, in a semisoft state, to a hot melt granulator where it is reduced to a free-flowing granulate.

2.11.2 Debaling and Initial Size Reduction

The first operation of the recycling process is the cleaning of foreign bodies. It requires a number of operating steps, the first being normally a debaling operation as the collected material, for transport reasons, is reduced into bales. Debaling is still often carried out manually. The reason often given for using a manual method is that the workman can also check the baled scrap for large pieces of foreign matter at the same time. However, there exist very efficient debalers for making the task automatic, and the best brands are equipped with specific devices designed in accordance to the composition of the bales to be loosened. Such factors as the forms of plastics items; the proportion of PE, PET, or PVC in the bales; the collection features; the container typology; and the share of foil plastics determine the type of debaler construction technique.

Two simple debaling services are a grab truck and a screw shredder. A grab truck can normally be used for bales of film. Sitting in his machine, the workman can break open the bales using three hydraulically operated grabs fitted to an extension arm. Checks for large pieces of foreign material can also be conducted. Using the same grabs, the loosened material can then be placed onto the feed conveyor fitted with a metal detector.

For bales containing individual items of scrap such as bottles and other hollow items, the use of a screw shredder offers many advantages. This machine is fitted with a very large feed hopper and can be fed directly with a large bucket loader or similar device. The feed material is reduced by a tearing process in the shredder between independently driven screw shafts fitted with shredding teeth. Screw shredders are manufactured with up to six adjacent shafts and are thus suitable for the feed of very large bales or a large number of bales at one time.

After debaling, the material passes through initial size reduction units to separation and selection operations. Most size reduction tasks can be performed by the following machines: shredder, cutter or guillotine, screw shredder, and granulator.

2.11.2.1 Shredder

Shredders have been in use for a long time in many sectors for the recovery of scrap. To a large extent they draw in the material automatically, and are suited for film, sheets, solid pieces, hollow items, cables, etc. The stresses caused by the tough and partly high-strength material are enormous. Extremely sturdy units designed for this technology are thus required for the treatment of plastics. It is important that the cutting shafts run at a suitable speed so that cutting and tearing processes occur.

Models of shredders are offered with from one to six cutting shafts. Machines with a capacity of many tones per hour are available. They are able to reduce complete bales of film fed by forklift without any difficulty. Also, hollow items such as rubbish bins and barrels can be reduced when a ram is fitted to the feed. Another field of application is the shredding of cables to allow the separation of plastics and metal. Generally, a trough fed by a forklift or conveyor is located above the shredder shafts. After the material is reduced to a practical size in the shredder, it is transported to subsequent process stages by means of a conveyor or other mechanical device.

2.11.2.2 Cutter or Guillotine

Some plastics scraps are not suited for initial reduction in a shredder described above. These include fibers, long pieces of material, rolled strips, and lumps of rubber. A guillotine is better for these applications.

The material is fed manually in a trough or on an open conveyor to the guillotine. The latter operates as opposed to the shredder, on a stroke principle. A cutter, usually hydraulically operated, is lowered from above to cut the material in slices of the desired thickness. The complete cutting process including the material feed is best operated on a programmed control.

2.11.2.3 Screw Shredder

The screw shredder mentioned earlier as a machine for debaling is also used for the initial size reduction of plastic items, and in particular when these are very voluminous and not too tough. It is suited for very large items or bundles of material and has the advantage of being fitted with a very large feed hopper that can be filled by bucket loader or similar device.

The machine is fitted with two shafts, rotating independently. It can be constructed, however, with up to six adjacent screw shafts, each shaft having its own drive via gears and electric motor. The shafts are equipped with shredder teeth for reducing the feed material in a crushing and tearing process. Not being a cutter, the machine is best suited for materials that can be broken or torn. When overloaded, a special control stops the respective shaft and switches it into reverse gear for a set time before returning to normal operation.

2.11.2.4 Granulators

Granulators can be seen as the most versatile size-reduction machine for the complete sector of plastics size reduction, and are used for the dry reduction of plastics. The machines used for this application are therefore designed to meet the special demands of job conditions, which are sturdy mechanical design, quick knife replacement, easy cleaning, and high capacity.

Since the reduction process is subject to the generation of a considerable amount of heat, it is necessary to water-cool parts of the machine or remove the heat with it. It is advantageous to fit the granulator with an open or semiopen rotor and a strong suction device to ensure that the grinding chamber is cooled intensively so that water-cooling is not required and the air is used at the same time for discharging the size-reduced material. All granulators should be equipped with a screen that can be easily removed. The screen opening determines the top size limitation of the size-reduced product. Material is fed to a feed hopper manually, or on a feed conveyor, screw, or similar device.

As an example, all granulators supplied by Herbold GmbH Maschinenfabrik (Meckescheim, Germany) have the following characteristics: (1) welded steel construction; (2) externally mounted bearings; (3) hinged two-piece housing with split point around the shaft; (4) easily replaceable screen; (5) double cross-cutting action; and (6) preadjustable knives. These characteristics offer a number of advantages, as explained below.

Due to the welded construction, the machine is resistant to extreme stresses caused by any foreign matter that may enter the granulator despite all precautionary measures taken, and fractures are avoided although the housing may be deformed. As the bearings are mounted outside the granulator housing, it is not possible for the feed material to enter the bearings or for grease to contaminate the material. The hinged housing allows easy and quick cleaning necessary when feed is changed and simplifies servicing and knife replacement. This is a significant advantage particularly in scrap recycling, where increased wear and more frequent knife replacement are to be reckoned with. For double cross-cutting action, all knives on the rotor are mounted at an inclined angle in a straight line to the rotor axis while all bed knives are set at the same angle but in an opposed inclined direction, also in a straight line. Complete sets of resharpened rotor- and bed-knives can be readjusted to the exact gap required between them to achieve the desired reducing action. This advantage, in conjunction with the good accessibility due to hinged housing, allows knife replacement to be carried out very quickly.

2.11.2.5 Fine Grinding

Fine grinding also offers solutions for the recycling of plastic scrap. Different types of machines are used, two common types being universal blast mills and disc pulverizers. In a universal blast mill, plastic scarp is reduced between the beater wings of a blast disc and a screen or the grooved grinding track of a grinding chamber. The hole size of the screen determines the fineness of the plastic powder. This usually has an upper size limitation of 500–800 μm.

In an impact disc pulverizer, which has much lower power consumption than the blast mill, the material is size reduced to an upper limit of about 800 μm between a fixed and a rotating disc. If a lower top size limitation is required, a screener may be used to return the coarse material continuously to the grind process. The finely ground powder is separated in a cyclone.

2.11.3 Cleaning and Selection

Cleaning, separation, and selection operations that usually follow the initial size reduction are determined by the type of recycling process to which the material is to be subjected. Basically there are two main recycling processes: recycle of heterogeneous plastics and recycle of selected polymers. The former process leads to the manufacture of extruded or injected products by direct reuse or in mixture with other components. The latter process consists of separation of the mix of collected plastics into homogeneous fractions, subjected to further processing that brings their characteristics and purity as near as possible to those of the original polymers.

The simplest method to perform the cleaning and selection operation consists of a selection platform where a number of trained sorters separate the different types of plastics on the basis of visual assessment. Though this is a hard and unpleasant job, the advantage of manual selection is that sorters operate to a degree of intelligence that the automatic equipment cannot reach. On the other hand, manual selection is, understandably, always liable to human error. To counter this problem, selection platforms are often equipped with detectors to check the quality of the selected material. These may be electronic appliances capable of recognizing, for example, PET in flux of PVC and vice versa, and detectors able to identify traces of metal overlooked during manual sorting, such as aluminum from caps and rings. The material manually selected and then electronically checked is therefore of best quality and can be sold at the maximum market price.

The most serious drawback of manual platforms lies in the high cost of labor and the need to manage a large number of workers when considerable quantities of material are to be sorted. Such drawbacks may be avoided by resorting to automatic platforms.

Automation is introduced at the stage of debaling. In order to obtain a product suitable for the recycling process, operations to remove undesired impurities must be carried out. The machines required are manifold and the necessity to employ them is related to the quality of the collected material. A few essential machines are: (1) rotary screen, by which parts of the desired dimension are sorted out, separating them from smaller and larger ones; (2) light-parts separation equipment, in which lighter parts such as films are separated by air blowing from the plastic material to be recycled; (3) heavy-parts separation equipment, in which heavy particles are separated and the operation is carried out by means of air that shifts the material selectively; and (4) aluminum rejection equipment, which normally consists of an electromagnetic drum placed in a suitable location on the train of operations. All such machines are preliminary to the stage of selection into homogeneous plastics fractions.

2.11.3.1 Dry Separation

The cleaning of plastics is often combined with the separation of other types of plastics and is performed by either dry or wet process depending on the quality of the collected material. A significant advantage of dry cleaning with air, as compared to wet cleaning, is that it has a lower power consumption. Loose adhering dirt—and this is the only type of sorting that can be removed in a dry process—is loosened and pulverized by the impact and rubbing caused during size reduction. The dirt, then as dust, can be separated by using an appropriate equipment. Examples of this are screen units and air stream separator.

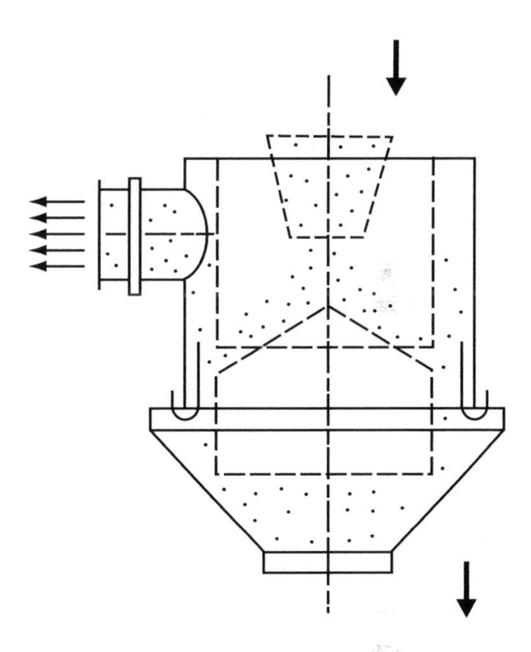

FIGURE 2.18 Schematic of air-stream separator.

The screen unit is the most economical means of removing the dirt. By selecting a screen with suitably sized openings, it is possible to minimize the amount of plastic discharged with the dust. Electrostatic charge can however cause too many dust particles to adhere to the plastic. The air separator (Figure 2.18) is then the more suitable device to use. Lighter-weight dust is carried out of the unit by the opposed directional air stream.

Mixed plastics, including various types of composite materials, that are to be dry-separated are first size-reduced (e.g., in a granulator) before the different constituents are separated. The separation is done in a process based on differences in material densities or shape and size of particles.

Screening may follow size reduction, depending on the material and the particle size distribution. In the screen unit, all of the mixed material is divided into two or three size fractions, e.g., 0–2, 2–4, and 4–6 mm. This is necessary since small heavy particles and large light pieces behave in the same manner, as do heavy flakes and light conical pieces. The separation based on density difference is easier or only then possible when all particles are of similar size.

Dry separation using air can be repeated several times, and the process is then classified as a cascade separator. A cascade separator, also known as a zigzag separator, uses an air stream passing through a rectangular zigzag channel from below (Figure 2.19). The material to be separated is fed to the top end of the channel into the air stream. The separation point is set by adjusting the air-flow rate. Material turbulence occurs in each section of the air stream channel, with lighter material being carried upward by air stream to discharge and heavy material moving downward from step to step. Fine particles are loosened from the larger ones and also from each other each time impact with side wall occurs. Cascade separators produce more efficient separation than single-stage units due to dispersion taking place at

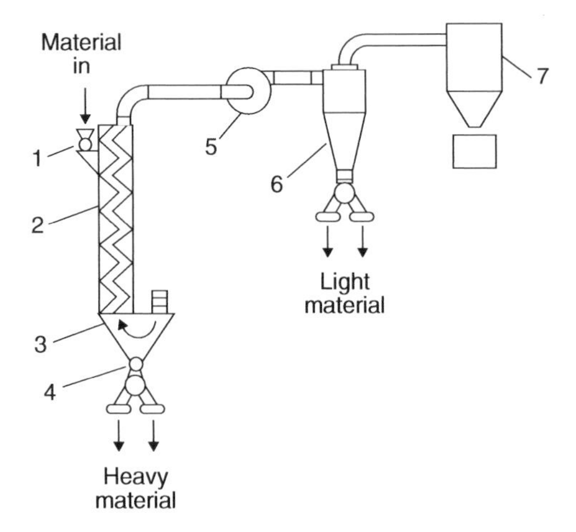

FIGURE 2.19 Schematic of air-stream (cascade) separator. 1, Gate valve; 2, cascading (zigzag) channel; 3, container with air suction; 4, gate valve; 5, blower; 6, cyclone; 7, filter with particle container.

each kink in the channel. Typical applications for the cascade separator are the separation of fibers and insulation film or foam and soft film.

The principle of operation of a fluidized air bed separator is shown in Figure 2.20. The material to be separated is carried uphill by the orbital vibration in the separation channel designed with a rectangular cross-section. An adjustable air stream is passed through the sieve surface in the channel where it lifts the material. The particles that are lifted higher in the air stream (that is, jump higher due to elasticity) flow downhill and are discharged from below. This type of fluidized air bed separation process can also be enhanced by using a multistage plant. A typical application for this unit is the separation of rubber from rigid thermoplastics or aluminum from plastics.

2.11.3.2 Wet Separation

Wet separation of plastics is a *microseparation method* in which a suspension medium is used to separate plastics with density higher or lower than the suspension medium. For example, water can be used as medium to separate PE from PVC or PET. In this case, special tanks are used in which various types of plastic flaks are mixed with water and then given a sufficient time to position themselves in the most suitable way according to their density. Materials are subsequently extracted separately from the top or bottom. This method is, however, not suitable for separating PVC from PET, because they have similar density.

Researchers at Rutgers University [70] studied a method of PVC microseparation from PET, by which PVC is subjected to a process of selective bulking that causes it to float. Such a method may be applied for separating small quantities of PVC from large quantities of PET, as normally is the case in the U.S., but is

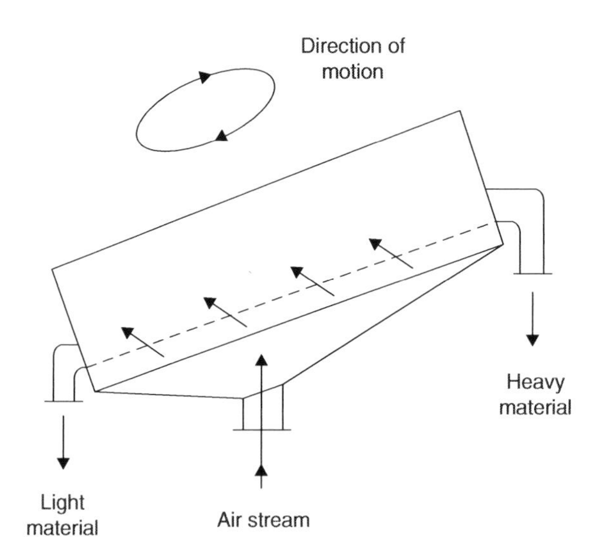

FIGURE 2.20 Schematic of vibrating air separator.

not suitable when PVC is a major component of the containers mix. Another wet method applicable to PVC and PET containers, previously reduced to flakes, is flotation with surface active agents.

2.11.3.3 Other Methods

Need for using microseparation techniques arises in critical situations where minute concentrations of identified contaminants adversely affect the post-consumer resin's usefulness in high-value end-uses. In many cases, microseparation, or the ability to sort resins by type, can be accomplished by air elutriation techniques (e.g., dry separation techniques and described above), wet separation techniques (such as sink/float tank technologies and hydrocyclones), magnets, electrostatic and electrodynamic methodologies, and optical scanners. Air aspiration and elutriation systems work well for separating light-density foams and films from denser reground plastics, while density-based methods are better for separating polyethylene from PET and denser resins.

Challenges to separate materials having similar densities, for example, PVC from PET regrind, or polypropylene from polyethylene, remain. Examples of microsorting techniques commercialized so far include electrostatic separation devices designed to sort by way of resin's conductivity, supercritical fluids which alter the separation fluid's density, froth flotation using the alteration of a liquid's surface tension to separate various solids, and chemical dissolution based on the difference in solubility of various plastics in selective solvents. Another development is a novel method to separate diverse resins by taking advantage of their differences in stick temperature.

BASF's Kali and Salz AG company, which has extensive experience in electrostatic separation of salts, employs its own electrostatic separation process (ESTA) to the separation of plastics. Using density separation, paper and plastic residues from labels and crowns are separated first. Then, following pretreatment with surface active substances designed to enhance the electrostatic properties, the homogeneously milled particles are charged electrically as they rub against each other. The extent of electrostatic charge depends on the plastic. The particles then fall through a high-tension field and are diverted at different angles depending on the charge, resulting in separation.

2.11.4 Resin Detectors: Type and Configuration

Detectors fall into four categories—x-ray, single-wavelength infrared (IR), full-spectrum IR, and color. The earliest automated systems used x-rays, which are still the most effective means of determining the presence of PVC. The chlorine atom in PVC emits a unique signal in the presence of x-rays by either x-ray transmission (XRT) or x-ray fluorescence (XRF). The XRT signal passes through the container, ignoring labels and other surface contaminants, and is capable of detecting a second container that may be stuck to the first. XRF, on the other hand, bounces off the surface of the container and is useful for finding any PVC, including labels and caps.

Systems for separating multiple types of plastics utilize a single wavelength of the near-infrared (NIR) spectrum. These systems work on the basis of a simple determination of opacity and separate the stream of mixed containers into clear (PET and PVC), translucent (HDPE and PP), and opaque (all pigmented and colored materials) streams.

The most sophisticated detectors, however, employ full spectrum NIR. Since all plastics absorb IR to different degrees and each resin has a unique "fingerprint," these detectors can accurately separate each of the resins. In later developments, filters for individual wavelengths are used for rapid identification and there is promise for even faster, lower-cost systems.

The color detectors are, in fact, very small cameras capable of identifying a number of colors. When combined with a resin-specific detector, they allows a variety of sorts based on both resin type and color.

Containers must be separated and presented to a detector in order to collect data on each unit. That information is then integrated via computer, and the container is tracked down the conveyor until it reaches the appropriate ejection point, where it is removed by a timed blast of air. The two primary techniques for presenting containers to a detector are full-conveyor or single-file systems. Containers delivered in a full-conveyor system can achieve higher throughput rates and are normally used to remove a single material from the mixture. A singulated stream, on the other hand, has lower throughputs, but allows sorting into a number of streams on a single pass.

Detectors are usually arranged in one of three configurations. The first is single detector/single container (Figure 2.21a): this is the simplest setup for singulated containers. As each container passes the detector, several readings are taken instantaneously and a decision is made by the computer. While usually accurate, this process is subject to error if the container has a large label blocking the signal and thus restricting data input to the computer. The second configuration is multiple detector/single container (Figure 2.21b): as each container passes the detector assembly, it is read by a number of detectors resulting in a more accurate reading. The third is multiple detector/multiple container (Figure 2.21c): This is the standard configuration for a mass-flow system and has detectors spaced to cover the width of the conveyor. When the target material is spotted, its position on the belt is noted and accordingly an ejector removes it before falling off at the end of the belt.

Further development in autosort technology is represented by particulate-sorting units capable of sorting by color. Applied in combination with the aforesaid resin detectors this facilitates autosorting according to both resin-type and color.

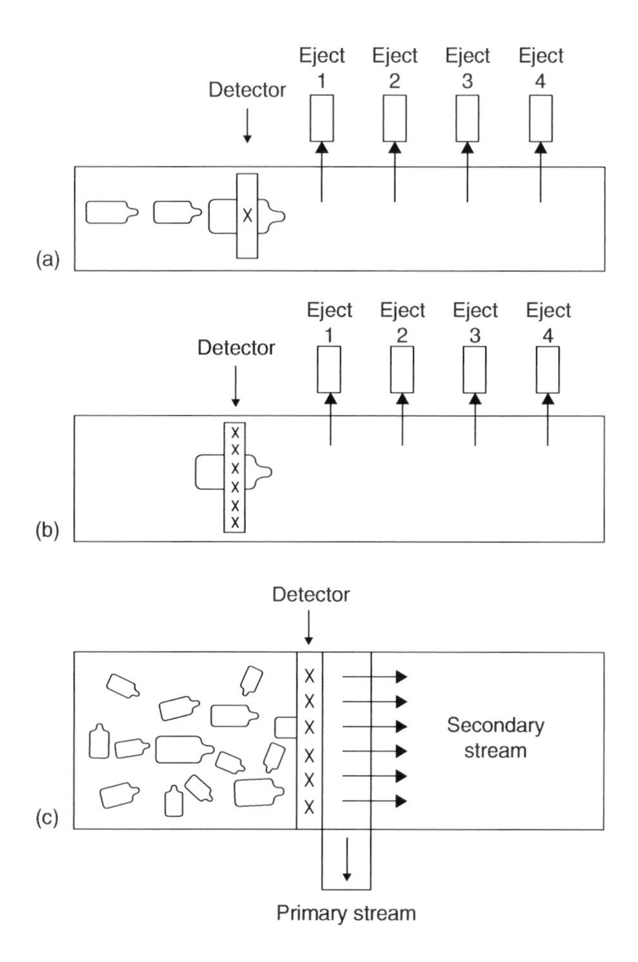

FIGURE 2.21 Typical separation and sorting setups using three main detector systems: (a) single detector/single sample; (b) multiple detector/single sample; and (c) multiple detector/multiple sample. (After Tomaszek, T. 1993. Automated Separation and Sort, *Modern Plastics*, 34–36 (November).)

2.11.5 Automatic Sortation

Most end-use markets of recycled plastics require that they be separated by resin type and color. For post-consumer bottles, all lids, caps, and closures should also be removed because they are often of different colors, and made of different resins than the bottle. The goal of any sorting process is to obtain the highest purity, consistency, and quantity of a particular consumer resin type. This ensures the highest end-use value for new products incorporating substantial amounts of the recycled resin.

In the widely used manual method of sorting, employees known as material handlers are stationed at predetermined locations alongside the sort conveyor to remove the desired bottles for recycling. All caps and closures still affixed to the bottles are manually removed by the handlers prior to or during sorting. These manual methods of sorting, however, face many challenges, some of which are economic and others environmental and aesthetic. Among the key hurdles are high cost of the labor-intensive process, exposure of employees to the residual household and industrial chemicals contained in some of the collected bottles, difficulty in sorting look-alike resins, and subjective material quality standards resulting from manual sorting. Automating the container sorting process to overcome these hurdles has thus been a major goal of the recycling industry.

An automatic separation process that includes various systems employing detectors currently available in the market is ideally suited to *macroseparation* or *macrosorting* (separation of plastic fractions before size reduction) when waste materials are still in the initial form (such as post-consumer bottles). Macrosorting is one of the fastest-growing segments of the plastics recycling industry. Although automated sorting, by resin type and color, can be accomplished after size reduction, most commercial automated-sorting machines are designed to sort plastics containers in whole form before size reduction. Some of the automated bottle-sorting systems are designed to sort certain bottles such as look-alikes, while others are designed to separate all plastics bottles by resin type and color. The most commonly encountered look-alike bottles are those fabricated from nonpigmented PVC and clear PET.

2.11.5.1 PVC/PET and Commingled Plastics Sortation

Both PVC and PET bottles are very recyclable. However, when the two types are received commingled, the reprocessor can experience quality deficiencies due to rheological incompatibilities between these two resins. This has been of special concern to the PET bottle reclaimer because, as the PET is heated to its processing temperature, trace amounts of PVC can cause severe deterioration in the quality of the reprocessed PET resin. Therefore the two resins must be separated prior to recycling.

Manual sorting of look-alike PVC and PET cannot meet the market's needed quality. Therefore new techniques have been engineered that will detect and separate bottles made from either of these two resins. The first detector was developed by Tecoplast in Casumaro, Ferrara-Italy, to separate PVC from PET. The application of this system resulted in the introduction of an automatic plant, processing drink plastic bottles using the AZZURRA machine. The Tecoplast detector consists of an x-ray source and a receiver that measures the bottle absorption while passing between the source and receiver. PVC has a higher absorption compared to the other plastic due to the presence of a chlorine atom. The value of PVC absorption, electronically processed through algorithms, makes it possible to detect its presence and consequently to eject bottle.

Another detector is employed to line up bottles using a suction robot. The aligning process enables bottles to be arranged in a suitable way for measuring transparency and color. Using this technology, Tecoplast developed the first optical detector capable of establishing the quality of PET, thus allowing the separation between clear PET and colored PET, besides the aforesaid separation from PVC. In a later development, Govoni's technology employed detectors performing these separations without alignment.

Other detectors have been developed in the U.S. A leading manufacturer of plastics sortation equipment, ASOMA Instrument Inc. (Austin, TX), developed a simple operator-friendly device for sensing the presence of chlorine atoms contained within the PVC resin. This sensor identifies PVC by x-ray fluorescence as bottles fall through a chute or move on a belt conveyor, at the end of which either the PVC or PET bottles are extracted—in most cases, by a burst of compressed air-jet actuated by the device, although mechanical devices have also been employed. The x-ray fluorescence sensitivity is so reliable that a 10 ms analysis is all that is needed to make proper selection. Single-unit systems have been developed that both detect and separate commingled bottles at production rates of 800–1200 lb/h, which corresponds to between two and three bottles per second. As is the case with most macroseparation devices, singulation (that is, alignment of bottles in a single file) is critical for this system to function at high production rates while maintaining rigorous quality standards. With x-ray detection, sorted PET streams having less than 50 ppm of PVC have been consistently produced, while, in contrast, manual separation generally results in PET with 2000 ppm PVC [71].

Another example is a system made by National Recovery Technologies (NRT) Inc. (Nashville, TN) [72]. This apparatus incorporates a proprietary electromagnetic screening process that also detects the presence of chlorine as found in PVC bottles. Once detected, PVC bottles are pneumatically jettisoned from the commingled bottle feed stream by a microprocessor-based air-blast system. The NRT technology permits bottles to be delivered to the unit in a mass sort concept in either crushed or whole form. This system does not require any special positioning or orientation of the bottles in order to achieve high efficiency rates.

An optical detector developed by Magnetic Separation Systems (MSS) Inc. (Nashville, TN) incorporates an optical sensing device with a transmission output range of 200–1500 nm to detect both the resin composition and the shape of the inspected container. Additionally, a video camera is employed to identify colored containers via computerized spectrographic matching. This information is also processed through a high-speed microprocessor that has the ability to perform algorithmic analyses and alarm the programmable logic controller (PLC) to actuate an ejection apparatus to sort the desired bottle. Finally, an x-ray fluorescence sensor is used to sort PVC bottles from the PET bottle fraction. The MSS detector system is thus designed to obtain separation of commingled plastics into homogeneous material fractions, including PVC, clear PET, colored PET, multicolored HDPE, and translucent HDPE.

A modular sorting system of MSS, BottleSort, Incorporates a sensory apparatus designed to detect and mechanically separate commingled plastic bottles in a process that includes several functions: debaling, screening, sensing, separation, and electronic control. Sensing is performed both optically and with x-ray fluorescence. Each BottleSort modular unit can process 1250 lb/h. Systems have been commercially installed incorporating four units having a combined capacity to sort 5000 lb of commingled bottles each hour.

A near-infrared spectrophotometry detector, developed by Automation Industrial Control (AIC) of Baltimore, MD, allows identification of resin type, such as PET, HDPE, PVC, PP, LDPE, and PS. The equipment is connected with another detector for color determination and the resulting data are processed by computer with highly sophisticated software. The equipment thus enables separation of a container mix into various components with a high degree of selectivity in regard to typology and color. This type of detector, however, requires material singularization and lining up.

The PolySort automatic plastic bottle sorting system introduced by AIC is designed to receive commingled plastic bottles in either baled or crushed form. At the heart of this sorting system is a sophisticated video camera and color monitor incorporating a strobe to detect and distinguish colors in the inspected plastic bottle. This optical scanning device interfaces with a computer to match the color of the bottle against a master. The detector is reputed to detect and match up to 16 million shades of colors. In addition, the system can be programmed to disregard labels on the bottles. Following color detection, a near-infrared detection system scans the single-bottle stream at a rate of approximately 3000 times a minute, to determine in less than 19 ms the primary resin found in each one. This is achieved by matching the interferogram produced by the bottle to a known master for each base resin as stored in the system.

Computations are coordinated through the use of a rotary pulse generator and sensing light curtain to impel the qualified bottle to a discharge chute located on the sort conveyor. Although the standard PolySort system is designed to detect and sort about 1500 lb of compacted bottles each hour, higher production rates can be obtained by feeding multiple lines from one debaler.

A point of consideration is the efficiency of separation that could nullify the high efficiency of a detection equipment. Detection must be unfailingly followed by rejection of the detected bottle, and in reality this does not always occur. Delivery systems in use today are effective but not 100% accurate. Most errors that occur with an autosort process are due to mechanical delivery errors rather than error related to detection. Therefore, a check on the operations of selection of the detected product is of primary importance. Detectors with 99% or greater efficiency, if installed in series and in number of at least two units, can bring the level of impurities within the limits required. For example, with a mix in the proportion of 90% PET and 10% PVC, two detectors with 99.5% efficiency placed in series enable one to obtain a level of impurities of 2.5 ppm, whereas with efficiency of 99% the value of residue is 10 ppm.

2.11.6 Recycle Installations

Special importance is generally attached to the techniques of electronic selection of homogeneous fractions, as discussed above, while disregarding the phase of regeneration of selected plastics. This would be justified if plastics articles were manufactured following criteria of perfect recyclability. However, such criteria are not yet universally adopted or followed. Therefore, an accurate response must be given, in any recycling plant design, to the problems posed by various elements that normally compose the item to be recycled. Referring to liquid containers, in general, these consist of, in addition to the body made of plastics, other foreign bodies that are to be removed. Such elements may be caps made of PE, PE with PVC gaskets, aluminum, labels of PVC or tacky paper with different types of glue, and residues and dirt that may have been added during the waste-collection phase.

Various operations, to be carried out in a specific sequence because of the problems posed by the type of material, are grinding, dry separation, and wet separation. Machineries to be installed for these operations are dictated by the typology and quality of recycle items.

Grinding is the first step following selection and requires attention and accuracy in design to ensure optimum homogeneity of the ground product. Several types of grinding equipment in common use have been described earlier.

The purpose of dry separation is the removal of a part in the dry phase. The process allows avoiding problems of dissolution in water and relevant contamination. An air flotation method is used for dry separation. Specially designed machines combining the effects of vibration and air flotation ensure separation of flakes with different specific weight. Such machines are very useful for removal of parts of labels that were freed by grounding. An extremely interesting application of this method is the separation of PVC labels from PET bottle flakes.

Residues are normally washed out of material flakes using a class of equipment that includes centrifugal cleaners, washing tanks, autoclaves, settling tanks, combined-action machines, scraping machines (mechanical friction), and centrifugal machines (for water separation). The construction details of the machinery and their installation according to a specific sequence are in the know-how of various manufacturers and very little is revealed. In view of complex problems posed by post-consumer plastics installations, it may, however, be said that it is not possible to expect miraculous results from key processes carried out in a single passage and therefore the efficiency is maximized by repetition of the same operation in more than one phase.

References

1. La Mantia, F. P. 1992. *Polym. Degrad. Stab.*, 37, 145.
2. La Mantia, F. P., Perone, C., and Bellio, E. 1993. Blends of polyethylenes and plastics waste. Processing and characterization, In *Recycling of Plastic Materials*, F. P. La Mantia, ed., pp. 131–36. ChemTec Publishing, Toronto-Scarborough, Canada.
3. Klason, C., Kubát, J., Mathiasson, A., Quist, M., and Skov, H. R. 1989. *Cellul. Chem. Technol.*, 23, 131.
4. Mathiasson, A., Klason, C., Kubát, J., and Skov, H. R. 1988. *Resour. Conserv. Recycl.*, 2, 57.

5. Boldizar, A., Klason, C., Kubát, J., Näslund, P., and Saha, P. 1987. *Inst. J. Polym. Mater.*, 11, 229.
6. Henstock, M. E. 1988. *Design for Recyclability.* The Institute of Metals, London, England.
7. Haylock, J. C., Addeo, A., and Hogan, A. J. 1990. Thermoplastic olefins for automotive soft interior trim. *SAE International Congress and Exposition,* Detroit, MI (Feb. 26–March 2, 1990).
8. Addeo, A. 1990. New materials for automotive interior. *22nd ISATA,* Florence, Italy (14–18 May, 1990).
9. Forcucci, F., Tomkins, D., and Romanini, D. 1990. Automotive interior design for recyclability. *22nd ISATA,* Florence, Italy (14–18 May, 1990).
10. Pfaff, R. 1990. Material recycling of polypropylene from automotive batteries—process and equipment, In *Second International Symposium,* J. H. L. Van Linden, ed., pp. 37–36. The Mineral, Metals and Materials Society, Warrendale, PA.
11. Basta, N. 1990. *Chem. Eng.,* 97, 37 (Nov. 1990).
12. Boeltcher, F. P. 1991. *ACS Polym. Prepr.,* 32, 2, 114.
13. Curto, D., Valenzen, A., and La Mantia, F. P. 1990. *J. Appl. Polym. Sci.,* 39, 865.
14. La Mantia, F. P. and Curto, D. 1992. *Polym. Degrad. Stab.,* 36, 131.
15. Ide, F. and Hasegawa, A. 1974. *J. Appl. Polym. Sci.,* 18, 963.
16. Cimmino, S., Cuppola, F., D'orazio, L., Greco, R., Maglio, G., Malinconico, M., Mancarella, C. et al. 1986. *Polymer,* 27, 1874.
17. Sinn, H., Kaminsky, W., and Janning, J. 1976. *Angew. Chem.,* 88, 737; Sinn, H., Kaminsky, W., and Janning, J. 1976. *Angew. Chem. Int. Ed. Engl.,* 15, 660.
18. Kaminsky, W. 1995. *Adv. Polym. Technol.,* 14, 4, 337.
19. Hirota, T. and Fagan, F. N. 1992. *Die Makromol. Chem., Macromol. Symp.,* 57, 161.
20. Paci, M. and La Mantia, F. P. 1998. *Polym. Degrad. Stab.,* 61, 417.
21. Scheirs, J. 1998. *Polymer Recycling Science, Technology and Application.* Wiley, New York.
22. Villain, F., Coudane, J., and Vert, M. 1995. *Polym. Degrad. Stab.,* 49, 393.
23. Datye, K. V., Raje, H., and Sharma, N. 1984. *Resour. Conserv.,* 11, 117.
24. De Winter, W. 1992. *Die Makromol. Chem., Macromol. Symp.,* 57, 253.
25. Hellemans, L., De Saedeleer, R., and Verheijen, J. 1997. U.S. Patent 4,008,048 to Agfa Gaevert.
26. Fisher, W. 1960. U.S. Patent 2,933,476 to Du Pont.
27. Gintis, D. 1992. *Die Makromol. Chem., Macromol. Symp.,* 57, 185.
28. Richard, R. 1991. *ACS Polym. Prepr.,* 32, 2, 144.
29. Fujita, A. A., Sato, S. M., and Murakami, M. M. 1986. U.S. Patent 4,609,680 to Toray Industries Inc.
30. Malik, A. I. and Most, E. 1978. U.S. Patent 4,078,143 to Du Pont.
31. Vaidya, U. R. and Nadkarni, V. M. 1988. *J. Appl. Polym. Sci.,* 35, 775; Vaidya, U. R. and Nadkarni, V. M. 1988. *J. Appl. Polym. Sci.,* 38, 1179.
32. Anon. 1991. *Eur. Chem. New,* 30 (Oct. 28, 1991).
33. Dabholkar, D. A. and Jain, M. U.K. Patent Application 2041916; C.A., 94: 209688r.
34. Thiele, U. 1989. *Kunststoffe,* 79, 11, 1192.
35. Adshiri, T., Sato, O., Machida, K., Saito, N., and Arai, K. 1997. *Kagaku Kogaku Ronbun.,* 23, 4, 505.
36. Sako, T., Okajima, I., Sugeta, T., Otake, K., Yoda, S., Takebayashi, Y., and Kamizawa, C. 2000. *Polym. J.,* 32, 178.
37. Genta, M., Tomoko, I., Sasaki, M., Goto, M., and Hirose, T. 2005. *Ind. Eng. Chem.,* 44, 3894.
38. Tokiwa, Y. and Suzuki, T. 1977. *Nature,* 270, 76.
39. Witt, V., Müller, R.-J., and Deckwer, W.-D. 1995. *J. Environ. Polym. Degrad.,* 3, 215.
40. Müller, R. J., Schrader, H., Profe, J., Dresler, K., and Deckwer, W.-D. 2005. *Macromol. Rapid Commun.,* 26, 1400.
41. Kaminsky, W. 1985. *Chem. Eng. Technol.,* 57, 9, 778.
42. Hagenbucher, A. 1990. *Kunststoffe,* 80, 4, 535.
43. Niemann, K. and Braun, U. 1992. *Plastverarbeiter,* 43, 1, 92.
44. Boeltcher, F. P. 1991. *ACM Polym. Prepr.,* 32, 2, 114.
45. Meister, B. and Schaper, H. 1990. *Kunststoffe,* 80, 11, 1260.

46. Müller, P. and Reiss, R. 1992. *Die Makromol. Chem., Macromol. Symp.*, 57, 175.
47. Grigat, E. 1978. *Kunststoffe*, 68, 5, 281.
48. Lentz, H. and Mormann, W. 1992. *Die. Makromol. Chem. Macromol. Symp.*, 57, 305.
49. La Mantia, F. P. ed. 1996. *Recycling of PVC and Mixed Plastic Waste*, pp. 265–36. ChemTec Publishing, Toronto-Scarborough, Canada.
50. Braun, D. 1992. *Die Makromol. Chem., Macromol. Symp.*, 57, 265.
51. Braun, D., Michel, A., and Sonderhof, D. 1981. *Eur. Polym. J.*, 17, 49.
52. Bonau, H. 1992. *Die Makromol. Chem., Macromol. Symp.*, 57, 243.
53. Dang, W., Kubouchi, M., Yamamoto, S., Sembokuya, H., and Tsuda, K. 2002. *Polymer*, 43, 2953.
54. Dang, W., Kubouchi, M., Sembokuya, H., and Tsuda, K. 2005. *Polymer*, 46, 1905.
55. Sembokuya, H., Yamamoto, S., Dang, W., Kubouchi, M., and Tsuda, K. 2002. *Network Polym.*, 23, 10.
56. Kriger, S. G. 1994. Recycling equipment and systems. *Mod. Plast.*, November E-47.
57. Vezzoli, A., Beretta, C. A., and Lamperti, M. 1993. Processing of mixed plastic waste, In *Recycling of Plastic Materials*, F. P. La Mantia, ed., pp. 219–36. ChemTec Publishing, Toronto-Scarborough, Canada.
58. Shenian, P. 1992. *Die Makromol. Chem., Macromol. Symp.*, 57, 219.
59. Dinger, P. W. 1994. Bottle reclaim systems. *Mod. Plast.*, A-46 (November 1994).
60. Hirota, T. and Fagan, F. N. 1992. *Die Makromol. Chem., Macromol. Symp.*, 57, 161.
61. Beckman, J. A., Crane, G., Kay, E. L., and Laman, J. R. 1974. *Rubber Chem. Technol.*, 47, 597.
62. Sperber, R. J. and Rosen, S. L. 1974. *Polym. Plast. Technol. Eng.*, 3, 2, 215.
63. Paul, J. 1986. In *Encylopedia of Polymer Science and Engineering*, Vol. 14, H. Mark, ed., p. 787. Wiley, New York..
64. Corley, B. and Radusch, H. J. 1998. *J. Macromol. Sci. Phys. B*, 37, 265.
65. Wu, S. 1985. *Polymer*, 26, 1855.
66. Swor, R. A., Jenson, L. W., and Budzol, M. 1980. *Rubber Chem. Technol.*, 53, 1215.
67. Tuchman, D. and Rosen, S. L. 1978. *J. Elastom. Plast.*, 10, 115.
68. Oliphant, K., Rajalingam, P., and Baker, W. E. 1993. Ground rubber tire-polymer composites, In *Recycling of Plastic Materials*, F. P. La Mantia, ed., ChemTec Publishing, Toronto-Scarborough, Canada.
69. Grigoryeva, O. P., Fainleib, A. M., Tolstov, A. L., Starostenko, O. M., and Lievana, E. 2005. *J. Appl. Polym. Sci.*, 95, 659.
70. Frankel, H. 1992. *Proceedings of Recycle 92*, Davos, Switzerland (April 7–10, 1992).
71. Tomaszek, T. 1993. Automated separation and sort. *Mod. Plast.*, 34–36 (November 1993).
72. Gorttesman, R. T. 1992. *Proceedings of Recycle 92*, Davos, Switzerland (April 7–10, 1992).

A1

Index of Trade Names and Suppliers of Foaming Agents

The following index of trade names and suppliers is based on a choice of representative foaming agents for thermoplastics. No claim is made for completeness. Detailed lists can be found in the source cited.

Chemical Foaming Agent	Trade Name and Supplier[a]	Processing Temperature Range, (°C)	Gas Yield (cm^3/g)	Recommended for
Azodicarbonamide (1,1′-azobisformamide)	Azofoam (Biddle Sawyer); Porofor (Miles); Kempore (Uniroyal); Ficel AC (Schering); Plastifoam (Plastics & Chem.); Cellcom (Plastics & Chem.); Unicell D (Dong Jin)	150–230	220	ABS, Acetal, Acrylic, EVA, HDPE, LDPE, PPO, PP, PS, HIPS, flexible PVC, TPE
Modified azodicarbonarnide (rigid PVC grade)	Kempore (Uniroyal); Plastifoam (Plastics & Chem.); Celogen AZRV (Uniroyal); Unicel DXRV (Dong Jin); Azofoam DS-1, DS-2 (Biddle Sawyer); Ficel AC2 (Schering)	150–200	155–230	PP, rigid PVC
Modified azodicarbonamide (flexible PVC open cell grade)	Azofoam F (Biddle Sawyer)	200–230	80–90	Flexible PVC
Dinitroso pentamethylene-tetramine	Opex 80 (Uniroyal); DNPT (Dong Jin); Mikrofine SSS (High Polymer Labs); Unicell GP9, GP3, GP42 (Dong Jin)	125–190	240	ABS, polyurethane, silicone, natural rubber, SBR

(continued)

Chemical Foaming Agent	Trade Name and Supplier[a]	Processing Temperature Range, (°C)	Gas Yield (cm^3/g)	Recommended for
4,4′-Oxybis (benzenesulfonyl) hydrazine	Cellcom OBSH-ASA2 (Plastics & Chem); Celogen OT (Uniroyal); Azofoam B-95 (Biddle Sawyer); Unicell OH (Dong Jin); Azocel OBSH (Fairmount); Mikrofine OBSH (High Polymer Labs)	120–190	120–125	EVA, LDPE, PS, flexible PVC
5-Phenyltetrazole	Expandex 5PT (Uniroyal); Unicell 5PT (Dong Jin); Plastifoam (Plastics & Chem.)	230–290	200	PFO, TPEs, PC, polysulfone, nylon, polyetherimide
p-Toluenesfulfonyl semicarbazide	Unicell TS (Dong Jin); Celogen RA (Uniroyal); Mikrofine TSSC (High Polymer Labs)	200–235	145	ABS, Acetal, Acrylic, EVA, HDPE, LDPE, PFO, PP, PS, HIPS, PVC, TPE
2,2′-Azobisiso butyronitrile	Mikrofine AZDM (High Polymer Labs)	90–115	125	Silicone rubber, semirigid PVC

[a]Names and addresses as follows: Biddle Sawyer Corp., 2 Penn Plaza, New York, NY 10121; Dong Jin (U.S.A.) Inc., 38 W. 32 St., Suite 902, New York, NY 10001; Fairmount Chemical Co., 117 Blanchard St., Newark, NJ 07105; High Polymer Labs, 803, Vishal Bhawan, 95 Nehru Place, New Delhi 110019, India; Miles Inc., 2603 W. Market St., Akron, OH 44313; Plastics & Chemicals Inc., P.O. Box 306, Cedar Grove, NJ 07009; Schering Berlin Polymers, 4868 Blazer Memorial Pkwy, Dublin, OH 43017; Uniroyal Chemical Co., World Headquarters, Middlebury, CT 06749.

Source: Modern Plastics, Mid-November 1994, p. C-73.

A2

Formulations of Polyurethane Foams

The formulations given below (in parts by weight) are three examples of typical formulations from the many available, which have been designed to meet specific requirements. Noteworthy points reading the formulations are explained. An example of formulation for reaction injection molding (RIM) and reinforced reaction injection molding (RRIM) is also given.

In order to achieve the desired balance of hydroxyl to isocyanate groups in a formulation, the *isocyanate index* is specified:

$$\textit{Isocyanate index} = \frac{\text{number of mole equivalents of isocyanate}}{\text{number of mole equivalents of polyols}} \times 100$$

If water in included in the formulation, this is also included in the mole equivalent of polyol.

A Typical Rigid Foam Formulation

Component A		
Polyol mixture containing a significant amount of triols and hex/octols to produce cross-linking		100
Catalyst 1	*N,N*-cyclohexylamine	0.3
Catalyst 2	*N,N*-dimethylethanolamine	0.3
Freon 11	($CFCI_3$)	50
Water		1
Surfactant	(A block copolymer of polyether and silicone)	1
Component B		
Impure liquid MDI	Equal in volume to component A and to give an isocyanate index near to 100	

Notes on Formulation A

1. In the formulation of the foam, the equipment perform better with approximately equal volumes of components A and B.
2. The MDI is of functionality around 2.2. This functionality together with aromatic nature of the MDI will tend to give rigid foams.
3. The isocyanate index is quoted as 100 to give conditions essentially producing urethance and urea group only.
4. Freon 11 blows to give closed cell structures whereas water produces open cells through carbon dioxide. This formulation is typical for a rigid foam for thermal insulation purposes. The heat build-up due to the reaction exotherm (about 80 kJ/mole for formation of urethane groups) is sufficiently dissipated through the open cell to avoid thermal degradation.

5. The catalyst combination allows a balance of reactions since catalyst 2 is specific to the water reaction. The values given are notional and depend on the polyols used.
6. A surfactant is used to stabilize the bubble structure.
7. It is likely that the product foam will have a density of 30 kg/m^3 which is roughly composed of 97% gas and 3% matrix by volume.

Formulation B A Typical flexible Foam Formulation

Component A	
Polyether polyol with long chains and overall low functionality	100
Water	4.5
Catalyst 1 (a tertiary amine)	0.15
Catalyst 2 (stannous octoate)	0.2
Freon 11	10
Silicone surfactant	1.3
Component B	
TDI (to index 112)	58.4

Notes on Formulation B

1. This is a standard foam grade formulation. Other flexible foam formulations are available for supersoft, high resilience and special grades.
2. The formulation is designed to give open cells.
3. The isocyanate has functionality of 2.0 and hence will not itself give cross-linking.

Formulation C A Typical flexible Foam Formation

Component A		
A flame retardant polyol		100
Catalyst 1,	DMP (tris-2,4,6-dimethylamionomethyl phenol)	3
Catalyst 2,	Sodium acetate: potassium acetate (1: 1) 33% w/w in ethane-1,2-diol	2
Freon 11		40
Surfactant	(silicone type)	1.5
Component B		
A liquid isocyanate based on MDI (to give index 200)		142.6

Notes on Formulation C

1. The high isocyanate index of catalyst system will promote the formation of isocyanate structures.
2. The higher heat stability of the isocyanate structures will be enhanced by a fire-retardant grade of polyol.

Formulation D A Typical RIM/RRIM Formulation

Component A		
Polyol mixture	Overall functionality 2.5, and containing primary hydroxyl terminated polyethers	100
Diol hardener	(1,4-Butanediol)	5
Soluble	(Soluble tin salt)	2

(continued)

Surfactant		1
Blowing agent	Freon 11	5
Milled glass fiber	(For RRIM)	5–30% of total
Component B		
Modified isocyanate based upon MDI		To match index 95–110

A Note on RIM and RRIM

RIM and RRIM require that a liquid isocyanate compound (component B) is effectively mixed with another liquid (component A) which contains polyols, catalysts, and other agents. Components A and B are metered and pumped into mixing heads where the two immiscible components are each turbulently broken down into small droplets surrounded by the other component phase. (The nozzles on the mixing head are usually 3 mm in diameter. This corresponds to a viscosity-pumping pressure relationship of 8 Pa s and 25 MPa, for example.) As the flow emerges from the mixing heads, the flow becomes laminar with phase separation (striation thickness) of about 10 μm. Reaction takes place at the interface during flow into the mold and subsequently in the mold until the products has sufficient strength to allow demolding.

The mixed liquids from the mixing head are forced (by pressure of liquid entering the mixing head) through a runner and gate and into the mold. It is generally required that the flow of the mixture should be laminar in the form of a film of about 1 mm thickness at a flow rate of about 2 m/s. Turbulent flow may cause air entrapment. The flow should be directed to the lower part of the mold to allow upward filling and any small amount of foaming to compensate for shrinkage (5%). The mold design should be such that air can escape through the parting lines.

Because of the low pressure (0.3 MPa) in the mold as compared to those encountered in thermoplastic injection molding (150 MPa) this process is suitable for the production of thin part with large surface areas. A significant application for this is in panel formation in the automotive industry, including fascias, door panels, spoilers, grills, and bumpers. While a rigid foam part with a flexural strength of 700 MPa would require a thickness of 7 mm, an RRIM part, because of the reinforcement, would allow a much smaller thickness (<3 mm) by virtue of the increase in flexural modulus.

A3

Formulations of Selected PVC Compounds

The formulations given below (in parts by weight) are example of typical formulations of PVC compounds from the many available, which have been designed to meet specific requirement of different applications.

Glass-Clear Food Packaging Film

PVC	100
Octyltin mercaptide	1–1.5
Glycerol ester	0.5–1.0
Ester wax	0.1–0.4

Furniture Film

PVC	100
Plasticizer	15–25
Epoxidized soybean oil	2–3
Liquid Ba/Cd stabilizer	1.5–2.0
UV absorber	0.2–0.4
Bis-amide wax	0.2–0.4
Processing aid	1
Pigment	0.5–10.0

Flexible Film for Outdoor Application

PVC	100
Plasticizer	40–80
Epoxy plasticizer	2–3
Tin carboxylate stabilizer	1–2.5
UV absorber	0.2–0.5
Oxidized polyethylene wax	0.2–0.4

Transparent Sheet for exterior Application

PVC	100
Butyltin mercaptide	2–2.5
Fatty alcohol	0.5–0.8
Fatty acid ester	0.5–0.8

(continued)

Polyethylene wax	0.1–0.2
Processing aid	1–2
UV absorber (benzotriazole type)	0.3–0.5

Aritificial Leather Cloth

Base coat	
PVC	100
Plasticizer	50–100
Epoxidized soybean oil	2–3
Barium/zinc stabilizer	1.5–2.0
Filler	0–20
Top coat	
PVC	100
Plasticizer	40–60
Epoxidized soybean oil	3–5
Barium/zinc stabilizer	1.5–2.0
UV absorber	0.2–0.3

Bottles for Food Packaging

PVC	100
Processing aid	0.5–1.0
Methyl-or octyltin mercaptide	1.0–1.5
Lubricant	0.5–1.5
Impact modifer	5–10

Potable Water Pipes

PVC	100
Methyltin mercaptide,[a] octyltin mercaptide,[a] butyltin mercaptide[a]	0.3–0.4
Calcium stearate	0.5–0.8
Solid paraffin wax	0.6–0.8
Polyethylene wax	0.1–0.2

[a]In compliance with specific national regulations.

Pressure Pipes and Conduits

PVC	100
Tribasic lead sulfate	0.5–1.0
Dibasic stearate	0.5–1.0
Calcium stearate	0.2–0.4
Stearic acid and/or paraffin wax	0.2–0.5

Calendered Floor Tiles

PVC	100
Plasticizer	30–70
Epoxidized soybean oil	2–5
Solid Ba/Cd stabilizer	1.5–2.5
Stearic acid	0.1–0.4
Chalk	50–100

Wire and Cable Insulation

PVC	100
Plasticizer	30–60
Tribasic lead sulfate	2–4
Normal lead stearate	0.5–1.5
Chalk	20–40
Stearic acid and/or paraffin wax	0–0.2

Windows Profiles

PVC	100
Impact modifier	6–12
Modified tin maleate stabilizer	2–2.5
Paraffin wax	0.6–1.2
Oxidized polyethylene wax	0.6–1.2
Antioxidant	0.1
UV absorber	0.2–0.4
Titanium dioxide	2–4
Chalk	10

Shoe

PVC	100
Plasticizer	50–80
Epoxidized soybean oil	2–3
Liquix Ba/Cd stabilizer	1.5–2.0
Stearic acid	0.2–0.4

A4

Formulations of Selected Rubber Compounds

Representative formulations (in parts by weight) of several rubber compounds and their normal curing conditions are given below.

Tread Compound for Passenger Tires

Smoked sheet	50
SBR 1712	70
Zinc oxide	5
Stearic acid	2
ISAF black	60
Softener	2
Antioxidant (diphenylamine-acetone condensate)	1.5
Accelerator (CBS)	1
Sulfur	2.2

Cure: 40 min. at 150°C.

Sidewall Compouond for Passenger Tires

Smoked sheet	50
SBR 1712	70
Zinc oxide	5
Stearic acid	2
HAF black	45
Softener	2
Antioxidant (diphenylamine-acetone condensate)	1.5
Accelerator (CBS)	1
Sulfur	2.2

Cure: 30 min. at 150°C.

Tube Compound for Car Tires

Butyl (Polysar 301)	100
p-Dinitroso benzene (Polyac)	0.15
FEF black	60
Mineral oil (butyl grade)	20
Zinc oxide	5
Stearic acid	2

(continued)

Accelerators	
MBT	1
TMT	1
Sulfur	1.5

Cure: 30 min. at 160°C.

Friction Compound for Conveyor Belts

Smoked sheet	100
Zinc oxide	5
Stearic acid	2
SRF black	10
Whiting/activated calcium carbonate	15
Tackifier/softener	5
Antioxidant	1
Accelerator (CBS)	0.6
Sulfur	2.5

Cure: 20 min. at 150°C.

Cover Compound for Conveyor Belts

Smoked sheet	50
SBR 1500	50
Zinc oxide	5
Stearic acid	2
Tackifier/softener	5
ISAF black	40
Antioxidant	1.5
Accelerator (CBS)	1.0
Sulfur	2.0

Cure: 20 min at 150°C.

Insulation Compound for Cables

Smoked sheet	100
Zinc oxide	20
China clay	30
Precipitated calcium carbonate	45
Paraffin wax	2
Stearic acid	0.5
Antioxidants	1.0
Accelerators	
DPG	0.5
MBTS	1.0
Sulfur	1.5

Cure: 15 min at 140°C.

Translucent Shoe Soling Compound

Pale crepe	100
Zinc oxide	3
Stearic acid	1
Precipitated silica	50
Paraffin wax	1

(continued)

Spindle oil	2
Diethylene glycol	2
Antioxidant (styrenated phenol)	1
Accelerators	
TMTM	0.5
ZDC	0.75
Sulfur	2.5

Cure: 7 min at 150°C.

Microcellular Shoe Soling

Smoked sheet	20
SBR 1500	20
SBR 1958	60
Peptizer	1
Microcellular crumb	60
Zinc oxide	5
Stearic acid	3–5
Paraffin wax	1
Mineral oil	10
Coumarone-indene resin	5
Styrenated phenol	1
Aluminum silicate	40
China clay	100
Blowing agent (DNPT)	5
Accelerators	
DPG	0.5
MBTS	1.0
Sulfur	2.5

Cure: 8 min at 150°C; oven stabilization at 100°C, 4 h.

A5

Commercial Polymer Blends and Alloys

The following index of trade names and suppliers is based on a choice of commercial polymer blends and alloys. No claim is made for completeness. Detailed lists can be found in the source cited.

Blend	Manufacturer	Composition	Reinforced[a]	Properties and Typical Uses
Alcryn	Du Pont	PO/EPDM	—	Processable, TPO
Arloy 1000	Arco Chem. Co.	PC/SMA	—	Automotive, medical
Arloy 2000	Arco Chem. Co.	SMA/PET	—	Food grade, transp.
Azloy	Azdel Inc.	PC/PBT	Yes	Automobile, electronics
Bayblend	Mobay/Bayer	PC/ABS	GF. Al	High impact strength, dimensional stability
Bexlov V	Du Pont	TPO blends	—	Automotive fascias
Bexloy W	Du Pont	Ionomeric alloys	—	Automotive bumper
Candon	Monsanto	SMA/ABS	—	Moldability. paintability
Celanex	Celanese	PBT/PET/Elast.	GF	Mechanical properties
Cycolac EHA	Borg-Warner	ABS/PC	—	Automotive applications
Cycoloy	Borg-Warner/Ube	ABS/PC or TPU	—	Heat, impact resistance
Cycoloy EHA	Borg-Warner	PC/ABS	—	Automotive applications
Denka HS	Denki Kagaku	ABS/PC	—	Automotive, electronic
Dia Alloy	Mitsubishi Rayon	ABS/PC	—	Automotive, electronics
Diacon	ICI	Acrylic/East.	—	Clear
Duraloy/Vandar	Celanese/Hoechst	POM/TPU or PBT	GF	Automotive, electronics
Durethan	Bayer AG	Nylon-6/PO or Elast.	—	Household appliances
Dynyl	Rohne-Poulenc	Nylon-6,6-modified block	GF	Low T impact, flex properties, sports goods
Ektar MB	Eastman Kodak	PCTG/PC or SMA	—	Electronic, appliances
Elemid	Borg-Warner	ABS/nylon	—	Auto, high T application
Envex	Rogers Corp.	PI/PTFE	—	Continuous-use $T =$ 225°C

(continued)

Blend	Manufacturer	Composition	Reinforced[a]	Properties and Typical Uses
Estane	B. F. Goodrich	TRU/SAN	—	Chemical/oil resistance
ETA Polymer	Republic Plast.	PO/EPDM	—	Automotive application
FerroFlex	Ferro Corp.	PP/EPDM	—	Automotive, electric
Fulton KL	LNP Corp.	POM/PTFE (20%)	—	Moving parts, automotive
Gafite/Celanex	GAF Corp./Hoechst	PBT/Elastomer	GF, micra	Electronics
Geloy XP 4001	G.E.	ASN/PC	—	Automotive
Geloy XP 2003	G.E.	ASN/PVC	—	Sidings, impact strength
Geloy SCC 1320	G.E.	ASA/PMMA	—	Gloss, surface hardness
Gemax	G.E.	PPE/PBT	Yes	Automotive applications
Geon	B. F. Goodrich	PVC/NBR	—	Coating, binding
Grilamid	EMS-Chem.	Nylon-12 aromatic-aliphatic PA	—	Eye-glass frames
Grilon	EMS-Chem. AG	Nylon-6/Elastomer	GF($\leq$30%)	Moldability, low T strength
Hostadur X	Hoechst AG	PBT/PET	GF	Computer, appliances
Hostaform	Hoechst AG	POM/TPU	MoS_2, PTFE	Impact strength
Hostalen	Hoechst AG	PP/EPDM	GF, talc	Automotive applications
Hostyren	Hoechst AG	PS/Elastomer	—	High impact PS
Hytrel	Du Pont	TPEs Elastomer	—	Blow molding
Idemitsu SC-250	Idemitsu Petrochem	PC/ABS,PES, Elastomer	—	Automotive, housings
Kelburon	DSM	PP/EPDM	—	Bumpers, suitcases
Keltan	DSM	PP/EPDM	—	Automotive parts
Koroseal	B. F. Goodrich	PVC/PVF	—	Linings
Kralastic	Uniroyal/Sumitomo	ABS/PVC	—	Moldability
Kraton D	Shell	SBS, SIS, SEP alloys	—	Automotive, sport
Kraton D2103	Shell	SBS/HIPS	—	Food containers
Kraton G	Shell	SEBS blends	—	Thermoplastic rubber
Krynac NV	Polysar Inc.	NBR/PVC (30–50%)	—	Weather, low T flex
Kydene	Rohm and Hass	PVC/PMMA	—	Thermoformable sheets
Kydex 100	Rohm and Hass	PVC/Acrylic	—	Thermoformable sheets
Lexan 100	G.E.	PC/PO	—	Electrical, housings
Lexan 500, 3000	G.E.	PC/PO	GF	Glass/metal replacement
Lomond	G.E.	PBT/SBS/ASA	—	Sporting, safety equip.
Luranyl	BASF	PPE/HIPS	GF, mineral	Housings, electronics
Makroblend PR	Bayer/Mobay	PC/PET or PBT	—	Bumpers
Makroblend	Bayer/Mobay	PBT/Elastomer	Yes	Automotive parts
Maranyl	ICI	Nylon-6 or 6,6/ Elastomar	GF, mineral	Sport, automotive parts
Merlon	Bayer/Mobay	PC/PO	—	Toughened PC
Mertex	Mobay	TPU blends	—	
Metamable	Teijin	PC/PMMA	—	Decorative use

(continued)

Blend	Manufacturer	Composition	Reinforced[a]	Properties and Typical Uses
Mindel A	Amoco	PSO/ABS	—	Hot water resistance
Mindel B	Amoco	PSO/PET	GF (40%)	High heat resistance
Minlon	Du Pont	Nylon-6,6/ionomer	Mineral	Low-T impact strength
Nipeon AL	Zeon Kasei	ABS/PVC (50%)	—	Good weatherability
Nipol	Nippon	NBR/PVC (30%)	—	Fuel hoses
Noryl	G.E.	PPE/HIPS	Yes	Processability, impact
Noryl FN	G.E.	PPE/HIPS	Foamable	Equipment housing
Noryl GEN	G.E.	PEF/HIPS	GF	Continuous-use T= 100°C
Noryl GTX (Noryl Plus)	G.E.	PA/PPE (30%)	Yes	Auto panels, wheels, fenders
Novalloy	Daicel	ABS/PC	—	Automotive, electrical
Novamate A	Mitsubishi	AAS/PC	GF (15%)	Electrical, electronic
Novamate B	Mitsubishi	ABS/PC	—	Automotive application
Novarex AM	Mitsubishi	PC/Elastomer	—	Car instrument panels
Novelen KR	BASF	PP/EPR	—	Self-supporting bumpers
Nydur	Bayer/Mobay	Nylon/Elastomer	GF (15%)	Low-T impact strength
Orgalloy	Atochem	Nylon-6/PP	—	Automobile body. Underhood
Pellethane	Dow	ABS/PTU	—	Automobile bumpers
Pocan S	Mobay/Bayer	PBT/Elastomer	—	Automotive applications
Polycomp	LNP Corp.	PPS or PET/	CF, GF	Bearings, cams, gears
Polyman 506	A. Schulman	PVC/ABS	—	Housings, appliances
Polyman 552	A. Schulman	SAN/PO	—	Recreational applications
Polysar	Polysar Inc.	PS/PB (4–8%)	—	Food containers
Prevex	Borg-Warner	PPE copal./HIPS	GF (≤30%)	Low-T impact strength
Pro-fax	Himont	PP/EPR	—	Automotive, houseware
Proloy	Borg-Warner	ABS/PC	—	Appliance housings
Propathane	ICI	PP/Elastomer	GF	Automotive applications
Pulse	Dow	PC/ABS (30%)	Yes	Auto panels, wheel covers
Rislan	Atochem	Nylon-6,6/PEBA	—	Sports goods
Riteflex BP	Hoechest Celanese	TPEs alloys	—	Golf carts, athletic shoes spoilers
Ronfalin	DSM	ABS/PC	—	Computer housings
Ronfaloy V	DSM	ABS/PVC	—	Business machines
Rynite SST	Du Pont	PET/Elastomer	GF (35%)	Automotive body parts
Santoprene	Monsanto	PP/EPDM	—	Thermoplastic rubbers
Saranex	Dow	PVDC/PE	—	Film applications
Selar	Du Pont	PA/PO	—	Blow molding
Styron XL	Dow	PS/Elastomer	—	Electronic
Technyl A	Rhone-Poulenc	Nylon-6,6/elastomer	—	Auto, recreational
Technyl B	Rohne-Poulenc	Nylon-6,6/Elastomer	GF (≤50%)	Mechanical and electrical properties

(continued)

Blend	Manufacturer	Composition	Reinforced[a]	Properties and Typical Uses
Telcar	Teknor Apex	PO/EPDM	—	Automotive application
Tenneco	Tenneco Polymers	PVC/EVA	—	Building industry
Terblend B	BASF	ABS/PS	Low-T impact	
Terblend S	BASF	ASA/PC	Yes	Auto, household applications
Texin	Mobay	PC/TPU	—	Thermoplastic rubber
Thermocomp PDX	LNP Corp.	PEEK/PTFE (20%)	—	Moving parts
Thermocomp	LNP Corp.	Nylon-6,6/Silicone	GF ($\leq$30%)	For injection molding
Torlon	Amoco	PAI/PTFE (3%)	Yes	Strength, thermal resistance
TPO 900	Reichold	PP/EPDM	—	Thermoplastic rubber
Triax 1120	Monsanto	Nylon-6/ABS	—	Impact heat, chemical resistance
Triax 2000	Monsanto	PC/ABS	—	Automotive market
Tribolon	Tribol. Ind. Inc.	PI/PTFE	—	Aerospace parts
Tribolon XT	Tribol. Ind. Inc.	PPS/PTFE	—	Moving parts
Tufrex VB	Mitsubishi Monsanto	ABS/PVC	—	Electronics housings
Ucardel P4174	G.E.	PSO/SAN	—	Single phase, transparent
Ultem	G.E	PEI/PC	$\leq$40%	
Ultrablend KR	BASF	PBTor PET/PC/ Elastomer	—	Bumpers, auto parts
Ultrablend S	BASF	PBT/ASA or SAN	GF ($\leq$30%)	Electronic, automotive
Ultramid/Terluran	BASF	Nylon/ABS	Mineral	Resistance to environmental stress cracking
Ultranyl	BASF	PPE/nylon	Yes	Automotive
Ultrason	BASF	PSO alloys	—	Electr., appliances
Valox 500 or 700	G.E.	PBT/PET or PBT/PC/ Elastomer	GF ($\leq$45%)	Dimensional stability
Vandar 8001	Hoechst-Celanese	PBT blend	Yes	Exterior automotive body panels
Vectra	Celanese	LCP blends	—	High-T mechanical properties
Vestoblend	Hüls	Nylon/PPE	—	Automotive applications
Vestolen	Hüls	PP/EPDM	—	Automotive, sport
Vestoran	Hüls	HIPS/PPE/Elastomer	—	Automotive
Victrex VKT	ICI	PEEK/PTFE (7.5–30%)	—	Moving parts, bearings
Victrex VST	ICI	PES/PTFE	—	Bearing applications
Xenoy 1000	G.F	PC/PBT (50%)	Yes	Low-T properties, car bumpers
Xycon	Amoco	TPEs/TPU	—	Bumper beams, electr.
Xyron 200	Asahi Chem. Ind.	PPE/HIPS	—	Office equipment
Xyron A	Asahi Chem. Ind.	PPE/nylon	—	Electric, automotive
Zytel 300, 400	Du Pont	Nylon-6,6/Ionomer	—	Tubing, cables

[a] *CF, carbon fiber; GF, glass fiber.*

Source: Utracki, L. A. 1989. Polymer Alloys and Blends: Thermodynamics and Rheology, Hanser Publishers, Munich/Vienna/New York.

Index

A

B

C

D

E

F

G

H

I

K

L

M

N

O

P

Q

R

S

T

U

V

W

X